PRACTICE WORKBOOK

On My Own

Harcourt Brace & Company
Orlando • Atlanta • Austin • Boston • San Francisco • Chicago • Dallas • New York • Toronto • London
http://www.hbschool.com

Printed in the United States of America

ISBN 0-15-307930-4

10 11 12 13 14 085 05 04 03 02 01

CONTENTS

CHAPTER 9 Place Value of Whole Numbers

CHAPTER 10 Comparing, Ordering, and Rounding Numbers

CHAPTER 11 Multiplication Facts 0–5

CHAPTER 12 Multiplication Facts Through 9

CHAPTER 13 Connecting Multiplication and Division

CHAPTER 14 Division Facts Through 9

CHAPTER 15 Collecting and Recording Data

CHAPTER 16 Representing Data

CHAPTER 17 Probability

CHAPTER 18 Classifying Plane and Solid Figures

CHAPTER 19 Making Figures Congruent

CHAPTER 20 Moving Plane Figures

CHAPTER 21 Fractions: Parts of a Whole

CHAPTER 22 Parts of a Group

CHAPTER 23 Decimals

CHAPTER 24 Measurement: Customary Units

CHAPTER 25 Measurement: Metric Units

CHAPTER 26 Measuring Plane Figures

CHAPTER 27 Multiplying by One-Digit Numbers

CHAPTER 28 Dividing by One-Digit Numbers

Name ______________________________

LESSON 1.1

Addition Strategies

Vocabulary

Complete. Choose **counting on** or **make a ten.**

1. When one of the addends is close to 10, first ______________________ and then add the rest.

2. When one of the addends is 1, 2, or 3, you can use ______________________ to find the sum.

Find the sum. Tell if you used *counting on* or *make a ten.*

3. $\begin{array}{r} 2 \\ +\ 7 \\ \hline \end{array}$

4. $\begin{array}{r} 9 \\ +\ 3 \\ \hline \end{array}$

5. $\begin{array}{r} 7 \\ +\ 8 \\ \hline \end{array}$

6. $\begin{array}{r} 4 \\ +\ 3 \\ \hline \end{array}$

______________ ______________ ______________ ______________

Find the sum.

7. $\begin{array}{r} 2 \\ +\ 9 \\ \hline \end{array}$

8. $\begin{array}{r} 6 \\ +\ 3 \\ \hline \end{array}$

9. $\begin{array}{r} 8 \\ +\ 4 \\ \hline \end{array}$

10. $\begin{array}{r} 3 \\ +\ 5 \\ \hline \end{array}$

11. $\begin{array}{r} 9 \\ +\ 7 \\ \hline \end{array}$

12. $4 + 6 =$ ______

13. $9 + 4 =$ ______

14. $3 + 7 =$ ______

15. $6 + 8 =$ ______

16. $4 + 7 =$ ______

17. $6 + 9 =$ ______

Mixed Applications

18. Jason walks 7 blocks to the library and 4 more blocks to the store. How many blocks does Jason walk in all?

____

19. Juan has a nickel, a dime, and a quarter. He spends 2 of the coins. What different amounts of money can he spend?

____

Name ______________________________

LESSON 1.2

More Addition Strategies

Vocabulary

Write the correct letter from Column 2.

1. $8 + 8 = ?$	____	**a.** doubles minus one
2. $8 + 9 = ?$ **Think:** $8 + 8 + 1$	____	**b.** doubles
3. $9 + 8 = ?$ **Think:** $9 + 9 - 1$	____	**c.** doubles plus one

---

Find the sum. Tell if you used *doubles, doubles plus one,* or *doubles minus one.*

4. $7 + 7$ ____________

5. $7 + 8$ ____________

6. $4 + 4$ ____________

7. $5 + 4$ ____________

8. $6 + 7$ ____________

9. $9 + 9$ ____________

10. $6 + 5$ ____________

11. $9 + 8$ ____________

Find the sum.

12. $9 + 9 =$ _____ **13.** $5 + 6 =$ _____ **14.** $5 + 5 =$ _____

15. $8 + 7 =$ _____ **16.** $3 + 4 =$ _____ **17.** $7 + 6 =$ _____

18. $6 + 6 =$ _____ **19.** $8 + 7 =$ _____ **20.** $3 + 3 =$ _____

Mixed Applications

21. A necklace has 7 blue beads and 6 red beads. How many beads are there in all?

____

22. Jill bought 4 corn muffins and 8 apple muffins. How many muffins did she buy in all?

____

Name ______________________________

Order and Zero

Find the sum.

1. $\begin{array}{r} 9 \\ +4 \\ \hline \end{array}$ $\begin{array}{r} 4 \\ +9 \\ \hline \end{array}$
2. $\begin{array}{r} 6 \\ +8 \\ \hline \end{array}$ $\begin{array}{r} 8 \\ +6 \\ \hline \end{array}$
3. $\begin{array}{r} 7 \\ +2 \\ \hline \end{array}$ $\begin{array}{r} 2 \\ +7 \\ \hline \end{array}$
4. $\begin{array}{r} 5 \\ +7 \\ \hline \end{array}$ $\begin{array}{r} 7 \\ +5 \\ \hline \end{array}$

5. $0 + 5 =$ ____
6. $9 + 0 =$ ____
7. $6 + 6 =$ ____
8. $8 + 0 =$ ____
9. $0 + 2 =$ ____
10. $0 + 7 =$ ____
11. $8 + 8 =$ ____
12. $0 + 4 =$ ____
13. $3 + 0 =$ ____

Find the sum. Use order in addition to write another addition fact. Example: $7 + 8 = 15$, so $8 + 7 = 15$.

14. $8 + 4 =$?, so ____ + ____ = ____
15. $3 + 7 =$?, so ____ + ____ = ____
16. $9 + 2 =$?, so ____ + ____ = ____
17. $6 + 5 =$?, so ____ + ____ = ____

Mixed Applications

18. Jennifer read 7 pages in the morning and 0 pages in the afternoon. How many pages did she read during the day?

____

19. Anne made 6 cookies. Carrie made one more cookie than Anne. How many cookies did they make in all?

____

20. Robbie and Betty each have the same number of pencils. Robbie has 4 new pencils and 8 used pencils. Betty has 8 new pencils. How many used pencils does Betty have?

____

21. A snail moved 3 inches in the morning and 7 inches in the afternoon. How many inches did the snail move in all?

____

Name ______________________

Problem-Solving Strategy

Make a Table

Make a table to solve.

1. List the addition number sentences that can be written using two sets of cards with the numbers 5, 6, and 7. Which of the number sentences have the same addends but show a different order?

____

____

____

2. Mr. Nolan kept track of the number of hot dogs he sold each day. On Monday, he sold 256 hot dogs and on Tuesday, 197 hot dogs. On Wednesday, he sold 286 hot dogs and on Thursday, 169 hot dogs. He sold 302 hot dogs on Friday. Order from least to greatest the number of hot dogs sold each day.

____

For Problems 3–4, use the table you made in Problem 2.

3. On which day did Mr. Nolan sell the most hot dogs?

____

4. On which days did Mr. Nolan sell fewer than 200 hot dogs?

____

Mixed Applications

Solve.

CHOOSE A STRATEGY

- **Guess and Check**
- **Look for a Pattern**
- **Make a Model**
- **Write a Number Sentence**

5. Dan and Andrew collected 13 shells in all. Dan collected one more shell than Andrew. How many shells did Dan collect?

____

6. Emily had 10 pieces of paper. She gave a piece of paper to each of her 3 friends. How many pieces of paper does Emily have left?

____

Name ______________________________

Subtraction Strategies

Vocabulary

Draw a line from the word to its description.

1. counting back

2. counting up

3. sames

4. zeros

a. When you subtract zero from a number, the difference is that number.

b. When you subtract 1, 2, or 3, count back to find the difference.

c. Begin with a smaller number and count up to a larger number.

d. When you subtract a number from the same number, the difference is zero.

Find the difference. Tell whether you used *counting back, counting up, sames,* or *zeros.*

5. $7 - 3$ ______

6. $9 - 0$ ______

7. $8 - 7$ ______

Find the difference.

8. $10 - 2$

9. $8 - 5$

10. $5 - 0$

11. $9 - 6$

12. $7 - 7$

Mixed Applications

13. Peter had 9 marbles. He lost 2 of them. How many marbles does Peter have left?

______

14. Li had 2 red marbles and 5 blue marbles. He found 6 blue marbles. How many blue marbles does Li have now?

______

Name ______________________________

Fact Families

Vocabulary

1. A ______________ is a set of related addition and subtraction number sentences using the same numbers.

2. Addition and subtraction are ______________ operations.

Find the sum. Write a related subtraction fact.

3. $8 + 4 =$ ____ 4. $5 + 9 =$ ____ 5. $3 + 8 =$ ____ 6. $7 + 4 =$ ____

Find the difference. Write a related addition fact.

7. $8 - 3 =$ ____ 8. $9 - 7 =$ ____ 9. $13 - 6 =$ ____ 10. $15 - 8 =$ ____

Write the missing number to complete each fact in the fact family.

11. $8 +$ ____ $= 13$ 12. $5 +$ ____ $= 13$ 13. $13 -$ ____ $= 8$ 14. $13 -$ ____ $= 5$

Write the fact family for each set of numbers.

15. 5, 2, 7

16. 9, 7, 16

Mixed Applications

17. Sue chose the numbers 9, 3, and 12 to write a fact family. Write her number sentences.

18. Tom has 15 tennis balls. He has 6 green balls, and the rest are yellow. How many of the balls are yellow?

Name ______________________________

LESSON 2.1

More Than Two Addends

Find the sum.

1. $(2 + 5) + 3 =$ ____ 2. $6 + (3 + 5) =$ ____ 3. $(4 + 5) + 9 =$ ____

4. $4 + (3 + 7) =$ ____ 5. $(4 + 3) + 6 =$ ____ 6. $(1 + 7) + 4 =$ ____

7. $2 + (6 + 6) =$ ____ 8. $(4 + 6) + 3 =$ ____ 9. $7 + (1 + 5) =$ ____

Group the addends. Then find the sum.

10. $2 + 4 + 6 =$ ____ 11. $4 + 1 + 9 =$ ____ 12. $3 + 5 + 8 =$ ____

13. $9 + 2 + 5 =$ ____ 14. $4 + 4 + 5 =$ ____ 15. $6 + 7 + 2 =$ ____

Mixed Applications

For Problems 16–18, use the table.

GERBIL SALES FOR ONE WEEK	
Day	Number of Gerbils Sold
Monday	5
Tuesday	4
Wednesday	3
Thursday	6
Friday	7

16. How many gerbils were sold in all on Monday, Tuesday, and Wednesday?

____

17. How many gerbils were sold in all on Thursday and Friday?

____

18. How many fewer gerbils were sold on Tuesday than on Friday?

____

19. Joe has 12 fish, and Ellen has 7 fish. How many more fish does Joe have than Ellen?

____

20. Mr. Jones has 5 dogs, 4 horses, and 7 cats. How many pets in all does Mr. Jones have?

____

21. Kate bought her dog when he was 8 months old. That was 3 months ago. How old is Kate's dog now?

____

Name ______________________

LESSON 2.2

Modeling Addition of Two-Digit Numbers

Use base-ten blocks. Write *yes* or *no* to tell if you need to regroup. Find the sum.

1. $\begin{array}{r} 14 \\ +25 \\ \hline \end{array}$ ______
2. $\begin{array}{r} 37 \\ +27 \\ \hline \end{array}$ ______
3. $\begin{array}{r} 15 \\ +16 \\ \hline \end{array}$ ______
4. $\begin{array}{r} 34 \\ +41 \\ \hline \end{array}$ ______
5. $\begin{array}{r} 31 \\ +62 \\ \hline \end{array}$ ______

Find the sum.

6. $\begin{array}{r} 43 \\ +24 \\ \hline \end{array}$
7. $\begin{array}{r} 26 \\ +34 \\ \hline \end{array}$
8. $\begin{array}{r} 16 \\ +29 \\ \hline \end{array}$
9. $\begin{array}{r} 41 \\ +38 \\ \hline \end{array}$
10. $\begin{array}{r} 62 \\ +28 \\ \hline \end{array}$
11. $\begin{array}{r} 81 \\ +70 \\ \hline \end{array}$
12. $\begin{array}{r} 63 \\ +41 \\ \hline \end{array}$
13. $\begin{array}{r} 48 \\ +68 \\ \hline \end{array}$
14. $\begin{array}{r} 39 \\ +81 \\ \hline \end{array}$
15. $\begin{array}{r} 98 \\ +58 \\ \hline \end{array}$

Mixed Applications

For Problems 16–18, use the table.

SIZE OF THIRD-GRADE CLASSES	
Class	**Number of Students**
Mr. Miller	22
Mrs. Foster	24
Ms. Lane	19

16. Ms. Lane's and Mr. Miller's classes are going on a field trip together. How many students are going on the field trip?

____

17. There are 20 students in Mrs. Foster's class today. How many students are absent?

____

18. There are 10 boys in Ms. Lane's class. How many girls are in her class?

____

Name ______________________________

LESSON 2.3

Adding Two-Digit Numbers

Find the sum.

1. $\begin{array}{r} 23 \\ +16 \\ \hline \end{array}$
2. $\begin{array}{r} 47 \\ +41 \\ \hline \end{array}$
3. $\begin{array}{r} 39 \\ +12 \\ \hline \end{array}$
4. $\begin{array}{r} 35 \\ +27 \\ \hline \end{array}$
5. $\begin{array}{r} 39 \\ +29 \\ \hline \end{array}$

6. $\begin{array}{r} 47 \\ +61 \\ \hline \end{array}$
7. $\begin{array}{r} 92 \\ +70 \\ \hline \end{array}$
8. $\begin{array}{r} 38 \\ +12 \\ \hline \end{array}$
9. $\begin{array}{r} 85 \\ +45 \\ \hline \end{array}$
10. $\begin{array}{r} 77 \\ +85 \\ \hline \end{array}$

Mixed Applications

For Problems 11–13, use the map.

11. Mrs. Fox drove from Ames to Canton. How many miles did she drive?

____

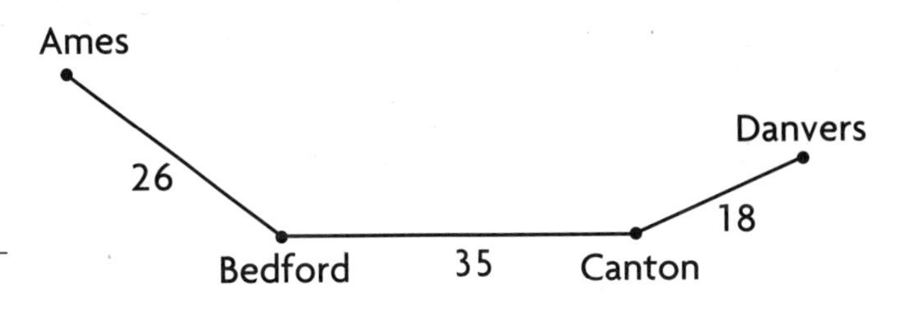

12. Mr. Clemente drove from Bedford to Danvers. How many miles did he drive?

____

13. Mrs. Ling drove from Canton to Danvers and back to Canton. How many miles did she drive?

____

14. Mr. Reed had 12 gallons of gas in his car. He used 4 gallons during the week. How many gallons of gas does he have left?

____

15. If you add me to myself, the sum is 16. What number am I?

____

Name ____________________

LESSON 2.4

Estimating Using a Number Line

Write the number that is halfway between the 2 tens. You may wish to use a number line.

1. 20 and 30 _____ 2. 70 and 80 _____ 3. 40 and 50 _____

Round each number to the nearest ten.

4. 34 _____ 5. 29 _____ 6. 14 _____ 7. 47 _____ 8. 35 _____

9. 64 _____ 10. 58 _____ 11. 65 _____ 12. 33 _____ 13. 76 _____

Estimate each sum or difference by rounding.

14. $23 + 71$

15. $93 - 43$

16. $84 - 19$

17. $69 + 12$

18. $32 + 48$

19. $67 + 17$

20. $65 - 19$

21. $52 + 34$

Mixed Applications

22. Carrie has 68 pennies in her piggy bank and 14 pennies in her wallet. About how many pennies does Carrie have in all?

____

23. To the nearest ten, a jar contains 60 marbles. What is the least number of marbles that can be in the jar? What is the greatest number?

____

24. Jane made a necklace with 7 red beads, 5 blue beads, and 3 white beads. How many beads did she use in all?

____

25. Chet collected 14 shells on Monday and 27 shells on Tuesday. Percy collected 39 shells on Monday and no shells on Tuesday. Who collected more shells?

____

Name ____________________

Choosing Addition or Subtraction

Tell if you need to *add* or *subtract*. Solve.

1. There are 38 boys and 42 girls taking swimming lessons at summer camp. How many children are taking swimming lessons in all?

   ____

2. The camp cook bought 35 boxes of cereal. During the week, the campers finished 31 boxes of cereal. How many boxes of cereal were left?

   ____

3. Peter hiked 2 miles on Monday, 4 miles on Wednesday, and 5 miles on Friday. How many miles did Peter hike in all?

   ____

4. On Monday morning, 36 campers went swimming, 14 campers went hiking, and 30 campers played games. How many more campers went swimming than hiking?

   ____

Mixed Applications

For Problems 5–7, use the pictured items.

5. Helen stopped at the gift shop on her way to a friend's house. She bought a memo pad and a pen. How much money did Helen spend?

   ____

6. Kito was told he could spend about 50¢ at the gift shop. Which items can Kito choose from?

   ____

7. Karen had 75¢. She bought a cup. How much money does Karen have left?

   ____

Name ____________________

Problem-Solving Strategy

Write a Number Sentence

Write a number sentence and solve.

1. It snowed on 4 days in December, 12 days in January, and 10 days in February. How many days did it snow in all?

____

2. Erin counted 12 birds at the bird feeder at 10 o'clock and 16 birds at noon. How many more birds were at the feeder at noon?

____

3. There were 25 tomatoes growing in Jared's garden. He picked 14 tomatoes. How many tomatoes were left in the garden?

____

4. Frank counted 5 red peppers, 11 green peppers, and 4 yellow peppers in his garden. How many peppers did he count in all?

____

Mixed Applications

Solve.

CHOOSE A STRATEGY

- **Make a Table**
- **Write a Number Sentence**
- **Act It Out**

5. Four students lined up from shortest to tallest. Mary is taller than Kim. Paul is shorter than Kim. Dave is taller than Mary. How were the students lined up?

____

6. Meg baked 36 cookies, Tom baked 60, Jon baked 72, Ann baked 48, and Ted baked 24. Who baked the most cookies? Who baked the fewest cookies?

____

7. Mr. Stewart bought 6 chocolate chip cookies, 3 sugar cookies, and 2 ginger cookies. How many cookies did he buy in all?

____

8. Ruth has 5 coins in her wallet. She has 21¢. What coins does she have?

____

Name ____________________

LESSON 3.1

Modeling Subtraction of Two-Digit Numbers

Use base-ten blocks. Write *yes* or *no* to tell if you need to regroup. Find the difference.

1. $\begin{array}{r} 25 \\ -14 \\ \hline \end{array}$ ______
2. $\begin{array}{r} 52 \\ -19 \\ \hline \end{array}$ ______
3. $\begin{array}{r} 36 \\ -17 \\ \hline \end{array}$ ______
4. $\begin{array}{r} 57 \\ -25 \\ \hline \end{array}$ ______
5. $\begin{array}{r} 82 \\ -65 \\ \hline \end{array}$ ______

Find the difference. You may wish to use base-ten blocks.

6. $\begin{array}{r} 75 \\ -37 \\ \hline \end{array}$
7. $\begin{array}{r} 24 \\ -16 \\ \hline \end{array}$
8. $\begin{array}{r} 38 \\ -25 \\ \hline \end{array}$
9. $\begin{array}{r} 82 \\ -26 \\ \hline \end{array}$
10. $\begin{array}{r} 43 \\ -36 \\ \hline \end{array}$
11. $\begin{array}{r} 64 \\ -29 \\ \hline \end{array}$
12. $\begin{array}{r} 25 \\ -15 \\ \hline \end{array}$
13. $\begin{array}{r} 73 \\ -45 \\ \hline \end{array}$
14. $\begin{array}{r} 54 \\ -19 \\ \hline \end{array}$
15. $\begin{array}{r} 36 \\ -24 \\ \hline \end{array}$
16. $\begin{array}{r} 47 \\ -28 \\ \hline \end{array}$
17. $\begin{array}{r} 98 \\ -44 \\ \hline \end{array}$
18. $\begin{array}{r} 36 \\ -28 \\ \hline \end{array}$
19. $\begin{array}{r} 52 \\ -18 \\ \hline \end{array}$
20. $\begin{array}{r} 65 \\ -18 \\ \hline \end{array}$
21. $\begin{array}{r} 61 \\ -27 \\ \hline \end{array}$
22. $\begin{array}{r} 47 \\ -30 \\ \hline \end{array}$
23. $\begin{array}{r} 96 \\ -66 \\ \hline \end{array}$
24. $\begin{array}{r} 77 \\ -28 \\ \hline \end{array}$
25. $\begin{array}{r} 56 \\ -38 \\ \hline \end{array}$

Mixed Applications

26. Dale had 24 math problems to solve. He now has 15 problems left to solve. How many problems has Dale already solved?

____

27. Kelly plants 24 seeds. Sara plants 16 seeds more than Kelly. How many seeds does Sara plant?

____

Name ______________________________

LESSON 3.2

Subtracting Two-Digit Numbers

Find the difference. Regroup if needed.

1. $\begin{array}{r} 37 \\ -18 \\ \hline \end{array}$ 2. $\begin{array}{r} 52 \\ -25 \\ \hline \end{array}$ 3. $\begin{array}{r} 47 \\ -31 \\ \hline \end{array}$ 4. $\begin{array}{r} 72 \\ -48 \\ \hline \end{array}$ 5. $\begin{array}{r} 65 \\ -39 \\ \hline \end{array}$

6. $\begin{array}{r} 42 \\ -31 \\ \hline \end{array}$ 7. $\begin{array}{r} 64 \\ -49 \\ \hline \end{array}$ 8. $\begin{array}{r} 82 \\ -47 \\ \hline \end{array}$ 9. $\begin{array}{r} 62 \\ -13 \\ \hline \end{array}$ 10. $\begin{array}{r} 93 \\ -67 \\ \hline \end{array}$

Mixed Applications

For Problems 11–13, use the table.

Trees in Park	
Pine	25
Oak	22
Maple	14
Birch	16

11. Jason counted the different kinds of trees in a park. How many trees did Jason count in all?

____

12. How many more pine trees are there than birch trees?

____

13. Of the 22 oak trees, 8 are white oaks, and the rest are black oaks. How many black oak trees are there?

____

14. Of the 25 students in Mike's class, 15 ride the bus, 4 ride in a car, and the rest walk to school. How many students walk to school?

____

15. Theresa left for school before Shelby. Scott left for school after Shelby, but before Manuel. In what order did the students leave for school?

____

____

Name ______________________

LESSON 3.3

Subtracting with Zeros

Find the difference.

1. $\begin{array}{r} 40 \\ -18 \\ \hline \end{array}$	2. $\begin{array}{r} 30 \\ -12 \\ \hline \end{array}$	3. $\begin{array}{r} 60 \\ -28 \\ \hline \end{array}$	4. $\begin{array}{r} 70 \\ -19 \\ \hline \end{array}$	5. $\begin{array}{r} 50 \\ -46 \\ \hline \end{array}$
6. $\begin{array}{r} 90 \\ -25 \\ \hline \end{array}$	7. $\begin{array}{r} 20 \\ -16 \\ \hline \end{array}$	8. $\begin{array}{r} 80 \\ -44 \\ \hline \end{array}$	9. $\begin{array}{r} 70 \\ -27 \\ \hline \end{array}$	10. $\begin{array}{r} 60 \\ -12 \\ \hline \end{array}$
11. $\begin{array}{r} 50 \\ -25 \\ \hline \end{array}$	12. $\begin{array}{r} 70 \\ -13 \\ \hline \end{array}$	13. $\begin{array}{r} 40 \\ -29 \\ \hline \end{array}$	14. $\begin{array}{r} 60 \\ -52 \\ \hline \end{array}$	15. $\begin{array}{r} 90 \\ -41 \\ \hline \end{array}$
16. $\begin{array}{r} 40 \\ -32 \\ \hline \end{array}$	17. $\begin{array}{r} 70 \\ -43 \\ \hline \end{array}$	18. $\begin{array}{r} 20 \\ -18 \\ \hline \end{array}$	19. $\begin{array}{r} 60 \\ -37 \\ \hline \end{array}$	20. $\begin{array}{r} 90 \\ -64 \\ \hline \end{array}$

Mixed Applications

21. Mr. Klein had 60 bottles of juice. He sells 18 bottles in the morning. How many bottles of juice does he have left?

____

22. Carl needs 50¢ to buy a bottle of juice. He has 38¢. How much more money does he need?

____

23. Mrs. Park drives 73 miles in the morning and 56 miles in the afternoon. How many miles does she drive in all?

____

24. Tony skates for 1 hour. He begins at 4:15. At what time will he finish skating?

____

Name ______________________

Practicing Subtraction

Find the difference.

1. $56 - 29$	2. $40 - 15$	3. $37 - 10$	4. $64 - 24$	5. $93 - 70$
6. $62 - 37$	7. $41 - 16$	8. $73 - 64$	9. $27 - 19$	10. $80 - 37$

Mixed Applications

For Problems 11–13, use the list.

11. Oscar collects stamps. He makes a list of the number of stamps he has from different countries. How many more stamps does Oscar have from the U.S. than from France?

____

12. How many stamps does Oscar have in all from the U.S. and Canada?

____

13. Oscar gives away 8 of his Canadian stamps. How many Canadian stamps does he have left?

____

14. Oscar begins working on his stamp collection at 3:15. He finishes 1 hour and 15 minutes later. At what time does Oscar finish?

____

15. Oscar buys a 32¢ stamp, a 10¢ stamp, and a 1¢ stamp. How much money does he spend in all?

____

Name ______________________________

Using Addition and Subtraction

Solve. For Problems 1–2, use the map.

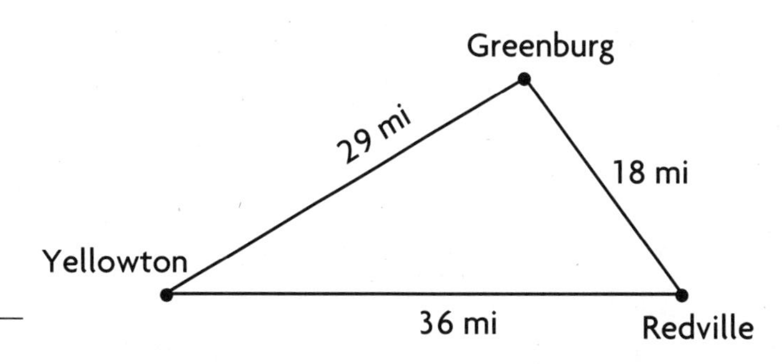

1. Mr. Harris drives from Yellowton to Redville. He then drives from Redville to Greenburg. How many miles does he drive in all?

____

2. How many miles closer is Greenburg to Yellowton than Redville to Yellowton?

____

Find the sum or difference.

3. $70 - 26$

4. $19 + 36$

5. $42 - 7$

6. $43 + 81$

7. $75 - 49$

8. $50 - 20$

9. $49 - 23$

10. $47 + 57$

11. $31 - 15$

12. $85 + 28$

Mixed Applications

13. The distance from Tammy's house to school is 23 blocks. Tammy has walked 18 blocks from her house toward school. How far away from school is she?

____

14. The dog walks around the edge of a square garden. The garden is 8 feet on each side. How many feet does the dog walk?

____

Name ______________________________

Problem-Solving Strategy

Work Backward

Work backward to solve.

1. Greta made some holiday cards yesterday. Today she made 3 more cards. She mailed 10 cards to friends. Greta now has 15 cards. How many cards did Greta make yesterday?

   ____

2. Steven picked some apples in the morning. He ate 2 apples for lunch. After lunch, he picked 14 more apples. Steven now has 30 apples in his basket. How many apples did he pick in the morning?

   ____

3. Nina had some money in her wallet. Then her father gave her 25¢. Nina spent 15¢. She now has 40¢ in her wallet. How much money did Nina have to begin with?

   ____

4. On Monday morning, Tom buys 8 baseball cards. On Tuesday, he sells 5 cards. He now has 45 baseball cards. How many cards did he have on Monday?

   ____

Mixed Applications

Solve.

CHOOSE A STRATEGY

- **Make a Table**
- **Work Backward**
- **Act It Out**
- **Write a Number Sentence**

5. Bill is fourth in a line of 10 people. Mary is in front of Bill. How many people are behind Mary?

   ____

6. Anne has 2 boxes of pens. There are 48 pens in each box. How many pens does Anne have in all?

   ____

7. Dave has 3 coins which have a total value of 36¢. What coins does Dave have?

   ____

8. Tim is thinking of a number. If you add 8 to the number and then subtract 10, you get 20. What is the number?

   ____

Name ______________________

Adding Three-Digit Numbers

Find the sum.

1. $356 + 228$
2. $149 + 227$
3. $657 + 155$
4. $494 + 369$
5. $364 + 465$
6. $648 + 173$
7. $649 + 348$
8. $146 + 594$
9. $247 + 453$
10. $152 + 688$
11. $384 + 165$
12. $473 + 437$
13. $349 + 449$
14. $147 + 366$
15. $869 + 131$

Mixed Applications

For Problems 16–17, use the table.

ELEMENTARY SCHOOL ENROLLMENT	
School	Number of Students
Davis	345
Lincoln	483
Lane	476
New Hope	372

16. How many students attend Davis and Lane elementary schools?

____

17. Which school has the most students? fewest students?

____

18. Ricardo found 11 new rocks for his collection during vacation. Now he has 47 rocks. How many rocks did he have before vacation?

____

19. Lisa read 73 pages on Tuesday and 46 pages on Wednesday. How many more pages did she read on Tuesday than on Wednesday?

____

Name ______________________________

LESSON 4.2

Adding More Than Two Addends

Find the sum.

1. 436 + 214 + 887

2. $3.50 + 4.49 + 3.69

3. 145 + 657 + 954

4. 329 + 416 + 741

5. $1.98 + 4.50 + 8.25

6. 608 + 652 + 654

7. 537 + 227 + 154

8. $5.16 + 5.25 + 4.64

Mixed Applications

For Problems 9–11, use the price list.

Price List	
Book	$3.50
Notebook	$1.65
Pen	$0.79
Markers	$2.25

9. Karen bought a book, a notebook, and a pen. How much did she spend for the items?

____

10. Hank has a $5 bill. Does he have enough money to buy markers, a notebook, and a pen?

____

11. Tanya has $0.50. How much more does she need to buy a pen?

____

12. Jill baked cookies in the morning. Her brother ate 4 of the cookies. In the afternoon, Jill baked 24 more cookies. She now has 44 cookies. How many cookies did Jill bake in the morning?

____

13. Jill put a batch of cookies into the oven at 2:35. She took them out at 2:45. For how many minutes did the cookies bake?

____

Name ____________________

Subtracting Three-Digit Numbers

Find the difference.

1. 354 − 148
2. 564 − 139
3. 942 − 817
4. 783 − 526
5. 647 − 435
6. 365 − 178
7. 635 − 145
8. 746 − 458
9. 852 − 459
10. 461 − 178
11. 461 − 275
12. 921 − 732
13. 437 − 128
14. 675 − 179
15. 724 − 536
16. 729 − 518
17. 436 − 297
18. 982 − 695
19. 514 − 226
20. 372 − 158

Mixed Applications

For Problems 21–23, use the table.

Attendance at School Play	
Friday	168
Saturday	314
Sunday	257

21. How many more people attended the play on Saturday than on Friday?

____

22. How many people attended the play on Friday, Saturday and Sunday?

____

23. There are 352 seats in the auditorium. How many seats were empty on Saturday?

____

Name ____________________

Problem-Solving Strategy

Guess and Check

Use guess and check to solve.

1. Two numbers have a sum of 39. Their difference is 11. What are the two numbers?

____

2. Two numbers have a sum of 22. Their difference is 4. What are the two numbers?

____

3. Gina traveled 450 miles to her grandmother's house in two days. She traveled 50 more miles on Saturday than on Sunday. How many miles did she travel on Saturday? on Sunday?

____

4. Maria practiced the recorder for 40 minutes on Saturday. She practiced 10 minutes less in the afternoon than in the morning. How many minutes did Maria practice in the morning? in the afternoon?

____

Mixed Applications

Solve.

CHOOSE A STRATEGY

- **Make a Table**
- **Act It Out**
- **Guess and Check**
- **Write a Number Sentence**

5. Jon visits 5 buildings in New York City. He records the heights of the buildings in feet as 625, 540, 505, 587, and 620. How tall is the tallest building Jon visits?

____

6. Molly has 6 coins that have a total value of 56¢. List the coins.

____

____

7. Shelia buys 76 cat stickers, 124 horse stickers, and 58 dog stickers. How many stickers does Shelia buy in all?

____

8. Jack had $5.10. He spent $3.95 for lunch. Does he have enough money left to buy ice cream for $1.25?

____

Name ______________________________

LESSON 4.4

Subtracting Across Zeros

Use base-ten blocks to find the difference.

1. $\begin{array}{r} 500 \\ -132 \\ \hline \end{array}$
2. $\begin{array}{r} 406 \\ -258 \\ \hline \end{array}$
3. $\begin{array}{r} 600 \\ -198 \\ \hline \end{array}$
4. $\begin{array}{r} 902 \\ -435 \\ \hline \end{array}$
5. $\begin{array}{r} 700 \\ -137 \\ \hline \end{array}$

6. $\begin{array}{r} 408 \\ -135 \\ \hline \end{array}$
7. $\begin{array}{r} 800 \\ -654 \\ \hline \end{array}$
8. $\begin{array}{r} 306 \\ -149 \\ \hline \end{array}$
9. $\begin{array}{r} 300 \\ -229 \\ \hline \end{array}$
10. $\begin{array}{r} 200 \\ -77 \\ \hline \end{array}$

Regroup. Write another name for each.

11. 400 ______________________

____

12. 507 ______________________

____

13. 206 ______________________

____

14. 800 ______________________

____

Mixed Applications

For Problems 15–16, use the map.

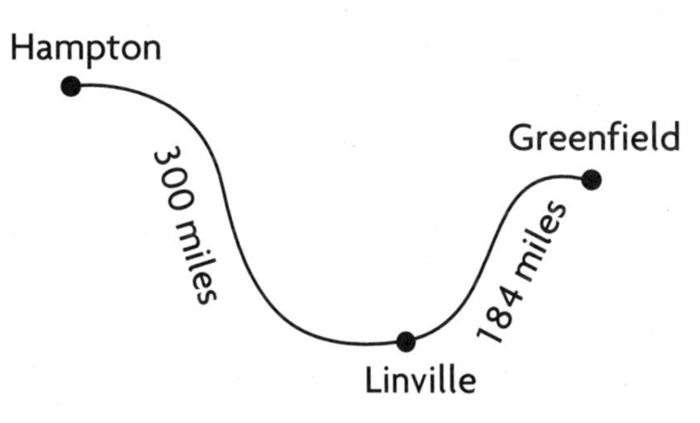

15. Jerry lives in Linville. How much closer does he live to Greenfield than to Hampton?

____

16. Mr. Crane is driving from Linville to Greenfield. He stops for lunch after driving 90 miles. How many more miles does he need to drive to reach Greenfield?

____

17. A farmer has 20 sheep and pigs. He has 6 fewer sheep than pigs. How many pigs does he have? **Hint:** Guess and Check.

____

Name ____________________

LESSON 4.5

More About Subtracting Across Zeros

Find the difference.

1. $400 - 135$
2. $700 - 471$
3. $900 - 198$
4. $300 - 264$
5. $800 - 123$
6. $\$4.00 - 1.25$
7. $500 - 481$
8. $\$7.00 - 3.50$
9. $600 - 488$
10. $200 - 65$

Mixed Applications

For Problems 11–13, use the table.

SOME MAJOR RIVERS IN NORTH AMERICA	
Name	Length
Hudson	306 miles
St. Lawrence	800 miles
Yellowstone	692 miles
Osage	500 miles

11. How much longer is the St. Lawrence River than the Hudson River?

____

12. Which river is about 200 miles longer than the Osage River?

____

13. The Alabama River is 71 miles shorter than the St. Lawrence River. How long is the Alabama River?

____

14. Dave bought a movie ticket for $3.75. He gave the salesperson a $10 bill. What was his change?

____

15. Sally bought a movie ticket for $3.75, a drink for $1.09, and popcorn for $2.25. How much money did she spend?

____

Name ______________________________

Understanding a Clock

Vocabulary

Draw a **minute hand** and an **hour hand** on the clock. Circle the hour hand.

1. 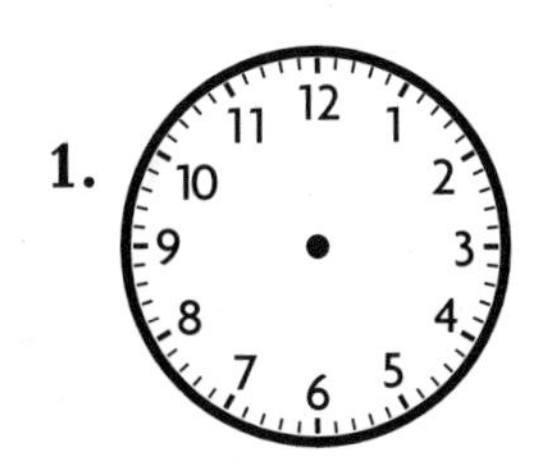

Tell how you would read each time. Write the time.

2. 

3. 

4. 

______________ ______________ ______________

______________ ______________ ______________

Write how many minute marks the minute hand has moved. Count by fives.

5. 

6. 

7. 

______________ ______________ ______________

Mixed Applications

8. Ana has 120 boxes of cookies to sell. So far she has sold 106 boxes. How many more boxes of cookies does she have to sell?

____

9. Alex goes to bed at 8:30 every night. He looks at the clock. The hour hand is pointing to the eight. The minute hand is pointing to the 3. Is it time for him to go to bed? Explain.

____

Name ______________________________

LESSON 5.2

Estimating Minutes and Hours

Write *more than* or *less than* for each.

1. It takes ____________ a minute to write your name.
2. It takes ____________ an hour to drive 100 miles.
3. It takes ____________ an hour to take a bath.
4. It takes ____________ a minute to write a poem.
5. It takes ____________ an hour to paint a house.

Circle the better estimate of time.

6. play a game of Go Fish	20 minutes or 20 hours
7. climb a mountain	3 minutes or 3 hours
8. eat an ice-cream cone	10 minutes or 10 hours
9. wash a car	30 minutes or 30 hours

Decide if the estimated time makes sense. Write *yes* or *no*.

10. It takes about 15 minutes to read a chapter. ____________
11. It takes about 3 hours to brush your teeth. ____________
12. It takes about 20 minutes to walk the dog. ____________
13. It takes about 1 hour to buy groceries for the week. ____________
14. It takes about 2 minutes to write a book report. ____________

Mixed Applications

15. Tony built a model airplane. Did Tony spend 3 minutes or 3 hours building the model?

____

16. There are 100 chairs in the cafeteria. There are 68 people sitting in chairs. How many chairs are empty?

____

Name ____________________

LESSON 5.3

Time to the Minute

Show each time on your clock. Draw a picture to show where the hour and minute hands are pointing.

1. 6:26

2. 8:02

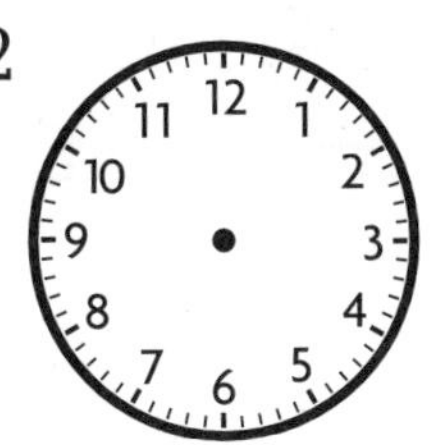

3. 7:48

4. 3:31

5. 12:53

6. 9:17

Write each time.

7.

____

8.

____

9.

____

Mixed Applications

10. The minute hand on a clock shows 28 minutes after the hour. How many minutes after the hour will it show in 12 minutes?

____

11. Julia's soccer team has played 3 games. The team scored 2 goals in the first game, 3 goals in the second game, and 4 goals in the third game. How many goals has the team scored?

____

Name ____________________

Time After the Hour

Write the two ways you would read each time.

1. 

2. 

3. 

Write each time.

4. 

5. 

6. 

7. 

Draw the hour hand where it should be.

8. 
11:14

9. 
2:52

10. 
7:02

Mixed Applications

11. On Tuesday, 137 children ate the school lunch. On Wednesday, 184 children ate the school lunch. How many more children ate the school lunch on Wednesday?

12. Ms. Roja's class goes to lunch at 12:00. Thirty minutes later they go outside for recess. Write the time recess starts. Tell where the hour hand is pointing.

Name ____________________

Problem-Solving Strategy

Act It Out

Act it out to solve.

1. Tika sits down to eat breakfast at seven o'clock. She leaves for school at 37 minutes after seven. Where are the hands on the clock when Tika leaves for school?

2. Mike starts his math homework at five o'clock. He is finished at 5:23. Where are the hands on the clock when Mike is finished?

3. Julie is sitting in the second row at the movies. Two people are sitting to her right. Five people are sitting to her left. How many people are sitting in the second row?

4. Kenny is playing a game with 10 counters. On his first turn, he loses 6 counters. On his next turn, he wins 3 counters. How many counters does he have now?

Mixed Applications

Solve.

CHOOSE A STRATEGY

- **Act It Out**
- **Write a Number Sentence**
- **Guess and Check**
- **Work Backward**

5. Iris went to the library after lunch. She was at the library for an hour. She left the library at 2:30. At what time did she get there?

6. Amy Lou had 67¢. She found a quarter on the sidewalk. How much money does Amy Lou have now?

Name ____________________

LESSON 6.1

Elapsed Time: Minutes and Hours

Vocabulary

Fill in the missing words.

1. ____________________ is the time that passes from the start of an activity to the end of that activity.

Use a clock with hands that move. Find the elapsed time.

2. start: 4:15
   end: 4:30

   ____

3. start: 5:30
   end: 5:45

   ____

4. start: 3:30
   end: 4:15

   ____

5. start: 4:45
   end: 5:15

   ____

Use a clock with hands that move. Find the ending time.

6. starting time: 4:15
   elapsed time: 30 minutes

   ____

7. starting time: 2:00
   elapsed time: 1 hour and 30 minutes

   ____

8. starting time: 7:30
   elapsed time: 45 minutes

   ____

9. starting time: 3:45
   elapsed time: 15 minutes

   ____

Mixed Applications

10. Toby's math class starts at 10:15. The math period is 45 minutes long. At what time is Toby's math class over?

    ____

11. Jessica's play rehearsal began at 3:15. The rehearsal ended at 4:30. How long was Jessica's rehearsal?

    ____

12. 14 girls and 9 boys are in Jan's class. How many more girls than boys are there?

    ____

13. Jerry drove 25 miles to the beach. Ann drove 19 miles. Who drove farther?

    ____

Name ________________________________

Using Time Schedules

Vocabulary

Fill in the missing word.

1. A(n) ______________ is a table that lists activities and the times they happen.

For Problems 2–5, use the schedule.

MONDAY ART CLASS SCHEDULE		
Class	Room	Time
Pottery	1	3:00–4:30
Painting	2	3:00–3:45
Drawing	2	3:45–4:30
Fabric crafts	3	2:45–3:45
Leather crafts	3	3:45–5:00

2. Which class meets in Room 1?

____

3. Which class lasts the longest?

____

4. If Felix takes painting and leather crafts, how much time will he spend in art classes on Monday?

____

5. Which classes are 45 minutes long?

____

Mixed Applications

6. How long is the morning recess in Ms. Reed's class?

____

MS. REED'S MORNING SCHEDULE	
Class	Time
Language Arts	8:15–9:45
Snack	9:45–10:00
Math	10:00–10:45
Recess	10:45–11:15
Social Studies	11:15–12:00

7. Chet practiced the piano for 25 minutes in the morning and 15 minutes in the afternoon. How long did he practice in all?

____

8. Susan had 308 marbles. She gave 25 marbles to her friend. How many marbles does Susan have left?

____

Name ______________________________

LESSON 6.3

Scheduling Time: Minutes and Hours

Complete the schedule.

CAMP WINDY MORNING SCHEDULE		
Activity	**Time**	**Elapsed Time**
Free play	9:00–9:30	________
Sports	________	1 hour 15 minutes
Snack	________	15 minutes
Crafts	11:00–12:00	________

For Problems 1–2, use the schedule you made.

1. How much longer is the sports period than the crafts period?

____

2. John arrived at camp at 10:00. How much time did he have to play sports?

____

Mixed Applications

For Problems 3–4, use the schedule.

CAMP WINDY AFTERNOON SCHEDULE	
Activity	**Time**
Lunch	12:00–12:30
Reading and games	12:30–1:30
Swimming	1:30–2:45
Free play	2:45–3:00

3. Which activity begins at 1:30?

____

4. Which activity in the afternoon is the longest?

____

5. There are 32 children at camp. On Tuesday, 18 children played kickball, and the others played tag. How many children played tag?

____

6. Molly made a necklace using 76 red, white, and blue beads. She used 36 red beads and 24 blue beads. How many white beads did she use?

____

Name ______________________

Scheduling Time: Days and Weeks

For Problems 1–2, use the schedule.

SCHOOL LUNCH SCHEDULE	
Day	**Food**
Monday	Hot dog
Tuesday	Pizza
Wednesday	Chicken
Thursday	Peanut butter and jelly sandwich
Friday	Hamburger

1. What day does the cafeteria serve chicken?

____

2. Elliot bought the school lunch on Tuesday. What did he have for lunch?

____

For Problems 3–6, use the schedule.

AFTER-SCHOOL SPORTS SCHEDULE		
Sport	**Day**	**Time**
Soccer	Mon	3:15–4:45
Baseball	Tue	3:30–5:00
Soccer	Wed	3:30–5:00
Baseball	Thu	3:15–4:30
Kickball	Fri	3:00–4:15

3. What days of the week does the soccer team practice?

____

4. How long does the baseball team practice on Thursdays?

____

5. Jessica plays soccer, and Jane plays kickball. What would be the best days for them to play together after school?

____

6. It takes Tony 15 minutes to walk home after kickball practice. What time does he get home?

____

Mixed Applications

7. Sam is reading a book that is 208 pages long. He has already read 79 pages. How many more pages does he have to read to finish the book?

____

8. Anne had 12 library books. She returned 5 books and checked out 7 more. How many library books does she have now?

____

Name ______________________________

LESSON 6.5

Elapsed Time: Days, Weeks, and Months

For Problems 1–3, use the calendar at the right.

July						
Sun	Mon	Tue	Wed	Thu	Fri	Sat
		1	2	3	4	5
6	7	8	9	10	11	12
13	14	15	16	17	18	19
20	21	22	23	24	25	26
27	28	29	30	31		

1. Tom is keeping Becky's hamsters at his house from July 13 to July 20. How many days is he keeping the hamsters? How many weeks?

____

2. Tom is feeding a cat from July 5 to July 19. How many days is he feeding it? How many weeks?

____

3. The Youngs are leaving on July 1 and will be away for 3 weeks. When will they return?

____

For Exercises 4–10, use your six-month calendar.
Write the date 4 weeks later.

4. January 6 ____________
5. March 15 ____________
6. April 2 ____________
7. April 29 ____________

Write the number of weeks.

8. from January 14 to February 25 ____________
9. from March 29 to April 19 ____________
10. from May 1 to June 5 ____________

Mixed Applications

For Problem 11, use your six-month calendar.

11. Li practiced for a play from April 24 to May 22. How many weeks did he practice?

____

12. Jim is 5 years older than his brother, who is 9 years old. How old is Jim?

____

Name ____________________

LESSON 6.5

Problem-Solving Strategy

Work Backward

Work backward to solve. For Problems 1–2, use your six-month calendar.

1. The date on Ben's calendar is March 29. Ben bought a kitten 5 days ago. The kitten was 9 weeks old when Ben bought it. On what date was the kitten born?

   ____

2. It is June 10. Linda has been working on an art project for 3 weeks. She bought the art supplies 4 days before she began the project. On what date did she buy the art supplies?

   ____

3. Carrie has been working on her homework for 30 minutes. Before that, she spent 15 minutes walking home from school. It is now 3:30. At what time did Carrie leave school?

   ____

4. It is now 9:30. Marco just spent 45 minutes raking leaves. Before that, he spent 30 minutes eating breakfast. At what time did he begin eating breakfast?

   ____

Mixed Applications

Solve.

CHOOSE A STRATEGY

- Make a Table
- Act It Out
- Guess and Check
- Write a Number Sentence

5. Jesse and Chet have 15 racing cars. Chet has 3 more cars than Jesse. How many cars does Chet have?

   ____

6. The Toy Corner ran a sale for 8 hours on Saturday and 6 hours on Sunday. For how many hours did the sale run?

   ____

7. Lucas spent 30 minutes playing. Then he read for 15 minutes. It is now 10:30. When did he begin playing?

   ____

Name ____________________

LESSON 7.1

Counting Bills and Coins

Count the money and write the amount.

1.   

____

2.     

____

3.  

____

4.   

____

Mixed Applications

For Problems 5–6, use the pictures.

5. Mark has one $1 bill, 2 quarters, and 1 dime. Which food item does he have enough money to buy?

____

6. Jim has exactly enough money to buy a hamburger. He has two $1 bills and 5 coins. What coins does he have?

____

7. On Monday, Mr. Harris sells 21 hot dogs, 36 slices of pizza, and 14 hamburgers. How many more hot dogs does he sell than hamburgers?

____

hot dog
$1.55

pizza
$1.75

hamburger
$2.09

Name ______________________________

Making Equivalent Sets

Vocabulary

Complete the sentence.

1. Sets that are ______________________ name the same amount.

Make an equivalent set with bills and coins. List how many of each bill and coin you used.

2.      

____

____

3.     

____

____

Make three equivalent sets for each amount. List how many of each bill and coin you used.

4. $1.60

____

____

____

5. $6.50

____

____

____

Mixed Applications

6. Josh has 7 dimes. Julia has the same amount in quarters and nickels. Julia has 6 coins. How many quarters and nickels does she have?

____

7. Tom has 46 pennies, 13 nickels, and 21 dimes. How many coins does he have in all?

____

Name ______________________________

LESSON 7.3

Comparing Amounts

Compare the amounts of money. Write the letter of the greater amount.

1. a.   

b.    

______

2. a.  

b.    

______

3. a.  

b.    

______

Mixed Applications

4. Frank has 7 quarters, 2 dimes, and 2 nickels. Leon has one $1 bill and 4 quarters. Who has more money?

____

5. How can you make 47¢ using the least number of coins?

____

____

6. Sheila begins climbing a mountain at 2:15. She reaches the top of the mountain at 3:00. How long does it take Sheila to reach the top of the mountain?

____

7. The Carlsons drive 300 miles on Monday and 279 miles on Tuesday. How many more miles do they drive on Monday than on Tuesday?

____

Name ____

Making Change

Use play money. List the coins you would get as change from a $1 bill.

1. $0.92 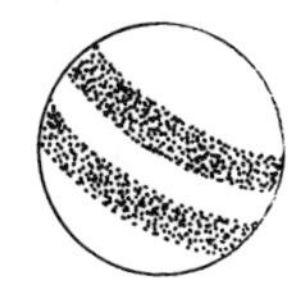

2. $0.35 

3. $0.59 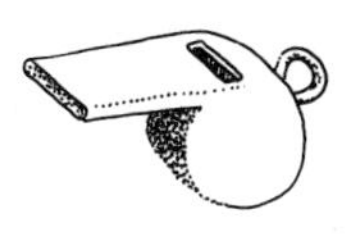

Use play money. List the coins and bills you would get as change.

	Amount Paid	Cost of Item	Change
4.	$1.00	$0.19	
5.	$5.00	$2.73	
6.	$6.00	$5.31	

Mixed Applications

7. Marco buys a book that costs $3.29. He gives the clerk $4.00. List the change he will get.

8. Luisa buys a bookmark for $0.39. She gives the clerk $1.00. What is the fewest number of coins Luisa can receive?

9. Mr. Murphy sells 9 bicycles on Monday, 8 bicycles on Tuesday, and no bicycles on Wednesday. How many bicycles does he sell in all during the 3 days?

10. On Monday, there were 30 balls in Mr. Murphy's store. On Friday, there are 12 balls left. How many balls did Mr. Murphy sell?

Name ______________________________

LESSON 7.5

Adding and Subtracting Money

Find the sum.

1. $\begin{array}{r} \$6.43 \\ +\ 2.15 \\ \hline \end{array}$

2. $\begin{array}{r} \$2.59 \\ +\ 1.37 \\ \hline \end{array}$

3. $\begin{array}{r} \$0.38 \\ +\ 5.24 \\ \hline \end{array}$

4. $\begin{array}{r} \$3.27 \\ +\ 2.06 \\ \hline \end{array}$

5. $\begin{array}{r} \$1.90 \\ +\ 2.64 \\ \hline \end{array}$

6. $\begin{array}{r} \$3.94 \\ +\ 2.75 \\ \hline \end{array}$

7. $\begin{array}{r} \$8.56 \\ +\ 4.03 \\ \hline \end{array}$

8. $\begin{array}{r} \$9.08 \\ +\ 1.35 \\ \hline \end{array}$

9. $\begin{array}{r} \$4.58 \\ +\ 2.67 \\ \hline \end{array}$

10. $\begin{array}{r} \$9.50 \\ +\ 7.68 \\ \hline \end{array}$

Find the difference.

11. $\begin{array}{r} \$5.63 \\ -\ 1.50 \\ \hline \end{array}$

12. $\begin{array}{r} \$4.93 \\ -\ 1.78 \\ \hline \end{array}$

13. $\begin{array}{r} \$6.55 \\ -\ 4.90 \\ \hline \end{array}$

14. $\begin{array}{r} \$4.02 \\ -\ 3.91 \\ \hline \end{array}$

15. $\begin{array}{r} \$3.50 \\ -\ 1.98 \\ \hline \end{array}$

16. $\begin{array}{r} \$5.00 \\ -\ 3.59 \\ \hline \end{array}$

17. $\begin{array}{r} \$4.50 \\ -\ 1.29 \\ \hline \end{array}$

18. $\begin{array}{r} \$10.00 \\ -\ 5.20 \\ \hline \end{array}$

19. $\begin{array}{r} \$20.00 \\ -\ 13.09 \\ \hline \end{array}$

20. $\begin{array}{r} \$3.80 \\ -\ 1.98 \\ \hline \end{array}$

Mixed Applications

For Problems 21–23, use the table.

FOOD FOR SALE	
Sandwich	$2.25
Drink	$0.79
Yogurt	$0.69
Salad	$3.75

21. David buys a sandwich for $2.25 and a drink for $0.79. How much does he spend in all?

____

22. How much more does a salad cost than a sandwich?

____

23. Dorinda pays $5.00 for yogurt and a salad. How much change will she get?

____

Name ______________________________

Problem-Solving Strategy

Write a Number Sentence

Write a number sentence to solve.

1. Jeff buys hamster food for \$2.25, a water bottle for \$4.00, and a food bowl for \$0.79. How much does Jeff spend in all? How much change will he get from \$10.00?

____

2. Jake buys a puzzle for \$2.25, a game for \$9.95, and a book for \$3.25. How much does Jake spend in all? How much change will he get from \$20.00?

____

3. Corina buys cloth for \$2.50, thread for \$0.89, and ribbon for \$1.55. How much does Corina spend in all? How much change will she get from \$5.00?

____

4. Elizabeth buys a cake mix for \$1.09, eggs for \$1.05, and candles for \$0.99. How much does Elizabeth spend in all? How much change will she get from \$5.00?

____

Mixed Applications

Solve.

CHOOSE A STRATEGY

- Make a Table
- Act It Out
- Guess and Check
- Work Backward

5. There are 26 students. There are 2 more boys than girls. How many girls are there?

____

6. It is 9:30. Recess begins in 45 minutes. At what time does recess begin?

____

7. Today's date on Lisa's calendar is July 24. She just spent one week at camp. Before that, she spent 3 days at her cousin's house. On what date did Lisa go to visit her cousin?

____

July						
Sun	Mon	Tue	Wed	Thu	Fri	Sat
		1	2	3	4	5
6	7	8	9	10	11	12
13	14	15	16	17	18	19
20	21	22	23	24	25	26
27	28	29	30	31		

Name ______________________________

LESSON 8.1

Ways to Use Numbers

Vocabulary

Complete.

1. Use a(an) ______________ number to tell how many.

2. Use a(an) ______________ number to show position or order.

For Exercises 3–6, use the words *HAVE A GOOD WEEKEND.* Answer each question. If your answer is a number, tell if it is a *cardinal* number or an *ordinal* number.

HAVE A GOOD WEEKEND!

3. In *HAVE,* in which position is the letter *V*?

____

4. In which word is the second letter *o*?

____

5. In which position is the third *E* in *WEEKEND*?

____

6. How many letters are in the second word?

____

Mixed Applications

The table shows how many books each class read in the Reading Marathon. For Problems 7–9, use the table.

READING MARATHON	
Class	**Number of Books**
Mr. Chen	134
Ms. Green	129
Mr. Gomez	113
Mrs. Anderson	109

7. Which class read the third greatest number of books?

____

8. In which place in the Reading Marathon is Ms. Green's class?

____

9. How many books in all did the four classes read?

____

Name ______________________________

Understanding 100's

Answer each question.

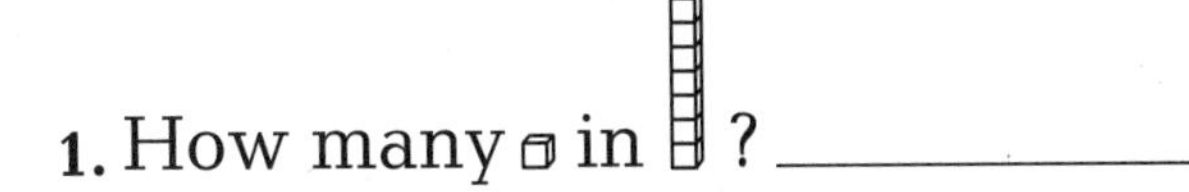

1. How many in ? ________

2. How many are equal to ? ________

3. How many in ? ________

4. How many are equal to ? ________

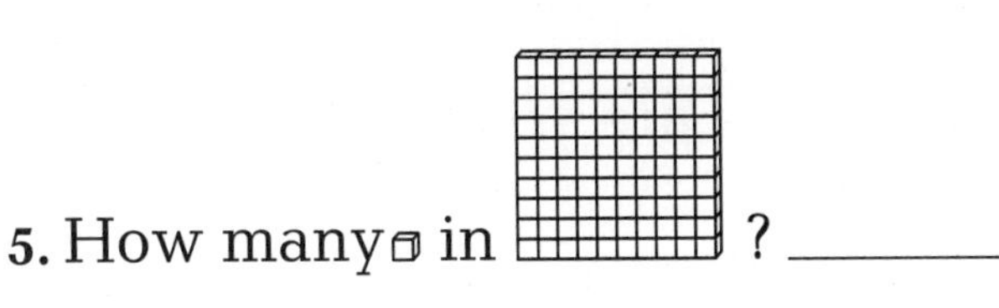

5. How many in ? ________

6. How many are equal to ? ________

Write *true* or *false*. Change words or numbers in the false sentences to make them true.

7. There are 100 ones in a ten.

____

8. There are 10 tens in a hundred. ________

9. There are 10 pennies in a dollar.

____

Mixed Applications

10. David has 40 pennies and 4 dimes. How many more dimes does he need to make a total of $1.00?

____

11. Nancy bought a pencil for $0.29. What is her change if she gives the clerk $1.00? What coins might she get?

____

____

Name ________________________________

LESSON 8.3

Number Patterns

Vocabulary

Complete.

1. Numbers ending with 0, 2, 4, 6, and 8 are ____________ numbers.

2. Numbers ending with 1, 3, 5, 7, and 9 are ____________ numbers.

Answer each question. Use a hundred chart to help you.

3. Skip-count by twos. Move 12 skips. Where are you? ______

4. Skip-count by threes. Move 5 skips. Where are you? ______

5. Skip-count by threes. Move 15 skips. Where are you? ______

6. Skip-count by fives. Move 9 skips. Where are you? ______

Tell whether the number is *odd* or *even*.

7. 34	8. 15	9. 82	10. 23	11. 19
________	________	________	________	________
12. 35	**13.** 81	**14.** 5	**15.** 89	**16.** 28
________	________	________	________	________

Mixed Applications

17. Mary had $0.36 when she got to school. On the way to school, she had found a penny and a dime. How much money did Mary have when she started out for school?

____

18. Write the numbers that are missing from this pattern.

30, 33, ____, 39, 42, ____, ____

Name ____________________

Patterns of Tens

Use patterns of tens to find the sum or difference.

1. $36 + 10 + 10 =$ ______
2. $49 + 10 + 10 =$ ______
3. $29 + 20 =$ ______
4. $47 + 30 =$ ______
5. $17 + 40 =$ ______
6. $47 - 30 =$ ______
7. $78 - 10 =$ ______
8. $95 - 10 - 10 =$ ______
9. $26 - 10 =$ ______
10. $54 - 30 =$ ______
11. $29 + 10 + 10 =$ ______
12. $52 + 40 =$ ______
13. $86 - 30 =$ ______
14. $39 + 40 =$ ______
15. Steve is counting backward by tens: 68, 58, 38, 28, 18, 8. Write a sentence to tell the mistake he made.

____

16. I am between 80 and 90. If you keep subtracting tens from me, you reach 4. What number am I?

____

Mixed Applications

For Problems 17–19, use the price list.

Item	Price
Pencil	\$0.05
Eraser	\$0.10
Ruler	\$0.20

17. Jackie bought 1 pencil and 3 erasers. How much money did she spend?

____

18. Tony bought 1 pencil, 1 eraser, and 1 ruler. How much money did he spend?

____

19. Peter had \$0.67. He bought 2 erasers. How much money does he have left?

____

Name ______________________________

Using Benchmark Numbers

Vocabulary

Complete.

1. ____________ numbers are useful numbers like 10, 25, 50, and 100 that help you see their relationship to other numbers.

Jar A — 10 beans

Jar B — 25 beans

Estimate the number of beans in each jar. Use Jars A and B as benchmarks.

2. 

12, 29, or 60 ______

3. 

14, 39, or 70 ______

4. 

25, 76, or 120 ______

For Exercises 5–6, write *more than* or *fewer than* for each. Use Jars A and B as benchmarks to help you estimate.

5. Are there more than or fewer than 10 beans in Jar X? ____________

6. Are there more than or fewer than 25 beans in Jar Y? ____________

Jar X

Jar Y

Mixed Applications

7. Tom needs 10 beans for science. Do you think there are enough beans in Jar X? Explain.

____

____

8. Barb needs 50 beans for an art project. She has 38 beans. How many more beans does she need?

____

Name ______________________________

LESSON 8.5

Problem-Solving Strategy

Make a Model

Make a model to solve.

1. Rose wants to fill a large basket with strawberries. She can put about 10 strawberries in a cup. How can Rose find out about how many strawberries she needs to fill the basket?

____

____

____

2. Dan is making paper airplanes for the school fair. He can make 1 airplane in about 5 minutes. About how many minutes will it take Dan to make 4 airplanes?

____

3. Jack is going to camp in one week. Today is July 16. When will Jack go to camp?

____

4. Mary had 7 dimes and 4 pennies. She earned 6 dimes. How much money does she have now?

____

Mixed Applications

Solve.

CHOOSE A STRATEGY

- **Guess and Check**
- **Act It Out**
- **Make a Model**
- **Work Backward**

5. Roger is saving acorns. He has a bagful. How can Roger find out about how many he has?

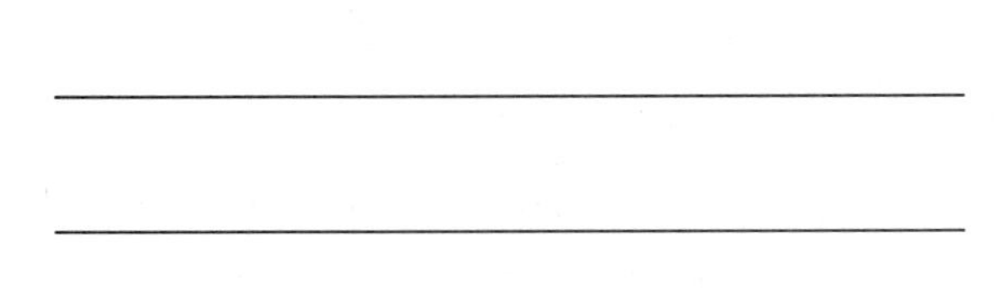

____

____

____

6. The clock shows the time Jill finished practicing the piano. She practiced for 45 minutes. At what time did Jill start practicing?

____

Name ______________________________

Value of a Digit

Vocabulary

Fill in the blank.

1. ______________ are the symbols 0, 1, 2, 3, 4, 5, 6, 7, 8, and 9.

Write the number represented by the base-ten blocks.

2. 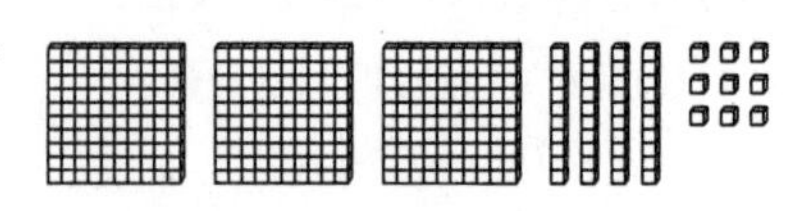

____

3. 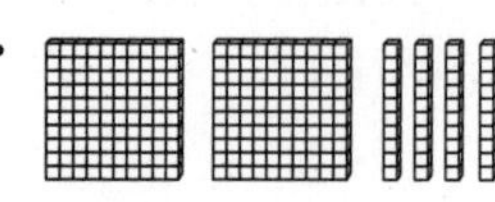

____

4. 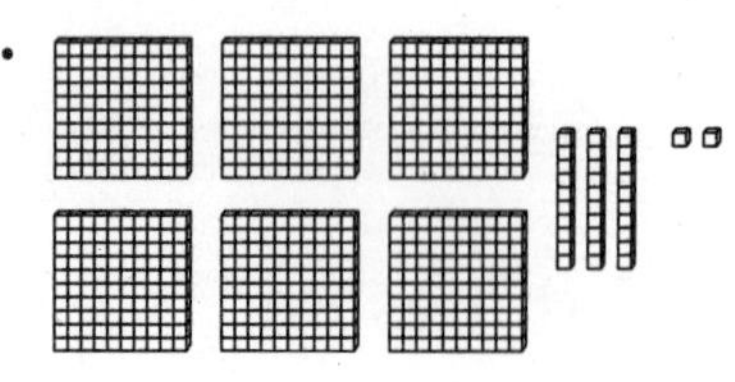

____

5. 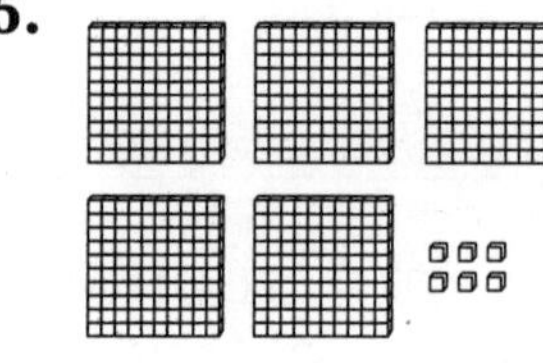

____

Write how many hundreds, tens, and ones.

6. 461 ______________________________

7. 302 ______________________________

8. 780 ______________________________

Write the value of the underlined digit.

9. 73<u>6</u> ______________

10. <u>3</u>41 ______________

11. 7<u>5</u>0 ______________

12. <u>4</u>08 ______________

Mixed Applications

13. Since April is the fourth month of the year, what month is May?

____

14. Jackie models 125 with beads. Bobbie gives her 3 more beads. What is the value of Jackie's beads now?

____

Name ______________________

LESSON 9.2

Understanding 1,000's

Page 1

5	10	15	20
25	30	35	40
45	50	55	60
65	70	75	80
85	90	95	100

Page 2

105	110	115	120
125	130	135	140
145	150	155	160
165	170	175	180
185	190	195	200

Page 3

205	210	215	220
225	230	235	240
245	250	255	260
265	270	275	280
285	290	295	300

Page 4

305	310	315	320
325	330	335	340
345	350	355	360
365	370	375	380
385	390	395	400

Page 5

405	410	415	420
425	430	435	440
445	450	455	460
465	470	475	480
485	490	495	500

Page 6

505	510	515	520
525	530	535	540
545	550	555	560
565	570	575	580
585	590	595	600

Page 7

605	610	615	620
625	630	635	640
645	650	655	660
665	670	675	680
685	690	695	700

Page 8

705	710	715	720
725	730	735	740
745	750	755	760
765	770	775	780
785	790	795	800

Page 9

805	810	815	820
825	830	835	840
845	850	855	860
865	870	875	880
885	890	895	900

Page 10

905	910	915	920
925	930	935	940
945	950	955	960
965	970	975	980
985	990	995	1000

Look at the grids above. Write the page number on which each number is found.

1. 92 ______ **2.** 225 ______ **3.** 626 ______

4. 422 ______ **5.** 167 ______ **6.** 968 ______

7. 899 ______ **8.** 351 ______ **9.** 735 ______

Mixed Applications

For the Problems 10–12, use the grids above.

10. Charlie lives at house number 650. On what page is 650?

____

11. Write four numbers that would be found on page 5.

____

12. Mary has 967 pennies. On what page is 967 found? ______

Name ______________________________

LESSON 9.3

Patterns of 100's and 1,000's

Use patterns of hundreds or thousands to find the sum or difference.

1. 710 + 200 __________
2. 335 + 100 __________
3. 421 − 200 __________
4. 563 − 400 __________
5. 806 + 100 __________
6. 436 + 200 __________
7. 4,271 + 1,000 __________
8. 6,702 − 1,000 __________
9. 5,326 + 2,000 __________
10. 8,763 − 8,000 __________
11. 2,196 − 1,000 __________
12. 4,305 + 3,000 __________

Mixed Applications

13. Tom had 250 rocks. He found 200 more. How many rocks does he have in all?

____

14. How long would it take to pour a glass of juice, 1 minute or one hour?

____

15. Julie had 548 books. She gave away 200 books. How many books does she have now?

____

16. Write a problem using what you know about how many books Julie has.

____

____

____

____

____

____

____

17. Abasi had 300 shells. He found 127 more. How many shells does he have now?

____

18. Shannon had 200 sheets of paper. She gave away 75 sheets. How many sheets of paper does she have now?

____

Understanding 10,000

Write each number.

1. 30,000 + 5,000 + 300 + 20 + 1 ______

2. 40,000 + 9,000 + 400 + 70 + 2 ______

3. 20,000 + 3,000 + 500 + 6 ______

4. 80,000 + 800 + 8 ______

5. 70,000 + 200 + 80 + 9 ______

6. 10,000 + 4,000 + 600 + 90 + 4 ______

7. sixty-one thousand, eight hundred thirty-one ______

8. forty-three thousand, five hundred forty-five ______

Write the value of the underlined digit.

9. 9<u>1</u>,643 ______

10. <u>3</u>6,955 ______

11. 72,<u>5</u>61 ______

12. 15,40<u>6</u> ______

13. <u>2</u>1,789 ______

14. 4<u>5</u>,632 ______

Mixed Applications

15. Suppose Mike received a check on Monday for $20,000, one on Tuesday for $9,000, one on Wednesday for $100, and one on Thursday for $28. How much money would Mike have in all? ______

16. The book Stacey is reading has fourteen thousand, two hundred twenty-two pages. Write this amount using digits. ______

Name ______________________________

Using Larger Numbers

Choose a benchmark of 1,000 or 10,000 to estimate or count each number.

1. 2,694 ____________
2. 1,456 ____________
3. 8,976 ____________
4. 42,965 ____________
5. 65,981 ____________
6. 15,426 ____________
7. 35,410 ____________
8. 4,325 ____________
9. 75,550 ____________

Write *yes* or *no* to tell if each number is large enough to be counted by using a benchmark of 1,000 or 10,000.

10. the number of desks in your classroom ____________
11. the number of seats in a football stadium ____________
12. the number of seats on a bus ____________

Mixed Applications

13. There were 8,456 people at the basketball game. 3,000 people left. How many people are still at the game?

    ____

14. Emily had band practice at 3:00. It lasted for 1 hour and 30 minutes. What time was band practice over?

    ____

15. The pizza store sells 3,120 pizzas each week. Which benchmark would be used to estimate the number of pizzas sold? Explain.

    ____

    ____

16. The music store has 44,261 CDs. Would a benchmark of 1,000 or 10,000 be used to estimate the number of CDs in the store? Explain.

    ____

    ____

Name ______________________________

LESSON 9.5

Problem-Solving Strategy

Use a Table

Use the table to solve.

1. Peggy's popcorn machine can make about 10,000 bags of popcorn a week. For which types of popcorn would it take more than a week to make all the bags?

____

2. One tub of kernels can make about 1,000 bags of popcorn. How many tubs of kernels does Peggy need to make caramel popcorn? Explain.

____

____

Peggy's Popcorn Palace	
Butter	15,460
Plain	11,326
Caramel	8,751
Unsalted	4,379
Honey nut	1,249

Mixed Applications

Solve.

CHOOSE A STRATEGY

- **Guess and Check**
- **Act It Out**
- **Make a Model**
- **Work Backward**

3. The sum of two numbers is 37. Their difference is 5. What are the two numbers?

____

4. Jim has $10.00. He buys a movie ticket for $4.00. What is his change?

____

5. There are 4 bags on the floor. Each bag has 8 soccer balls in it. How many soccer balls are there in all?

____

6. Joan had 156 stickers. She gave some away. Now she has 56 stickers. How many stickers did she give away?

____

Name ____________________

Comparing Numbers

Draw base-ten blocks to show your models. Circle the picture that shows the greater number.

1. 256 and 266
2. 50 and 51
3. 136 and 138
4. 161 and 116
5. 355 and 365
6. 43 and 44

Draw base-ten blocks to solve.

7. Steven has 4 hundreds, 5 tens, and 6 ones. Lenny has 5 hundreds, 3 tens, and 6 ones. Who has the greater number?

____

8. Kristen modeled 4 hundreds, 2 tens, and 1 one on page A. She has 3 hundreds, 0 tens, and 9 ones on page B. Which page shows the greater number?

____

Mixed Applications

9. Casey has 351 ones. He wants to trade some for hundreds and tens. How many hundreds and tens can he get? How many ones will be left?

____

____

10. Annie is having a party. She invites 18 of her friends, but 7 of them can't come. How many of Annie's friends will be at the party?

____

Name ______________________________

LESSON 10.2

More About Comparing Numbers

Compare the numbers. Write <, >, or = in each ◯.

1.

T	O
4	5

T	O
7	4

45 ◯ 74

2.

T	O
6	3

T	O
8	3

63 ◯ 83

3.

T	O
2	2

T	O
2	4

22 ◯ 24

4.

H	T	O
4	2	1

H	T	O
4	1	1

421 ◯ 411

5.

H	T	O
2	4	3

H	T	O
3	4	2

243 ◯ 342

6.

H	T	O
1	2	0

H	T	O
1	3	0

120 ◯ 130

7. 25 ◯ 32
8. 54 ◯ 69
9. 45 ◯ 44
10. 13 ◯ 14

11. 254 ◯ 255
12. 451 ◯ 448
13. 621 ◯ 612
14. 789 ◯ 790

Mixed Applications

15. Jillian spent $52 on a dress. Melinda spent $55 on a dress. Who spent more?

____

16. Kevin has 210 stamps. Eric has 212 stamps. Who has the greater number of stamps?

____

17. Lisa lives at house number 125. Jodi lives at house number 135. Who lives at the house with the higher number?

____

18. Joey ate 12 cookies on Wednesday and 15 cookies on Thursday. On which day did Joey eat more cookies?

____

19. Lisa spent $12.00 on movie tickets, $3.00 on popcorn, and $2.50 for a soda. Who spent more, Lisa or her friend who spent $17.75?

____

20. Cedric's lunch costs $2.30. He has a dollar bill, 3 quarters, and 5 dimes. Will he be able to pay his bill?

____

Name ______________________________

LESSON 10.3

Ordering Numbers

Write the numbers in order from least to greatest. Use the number lines to help you.

1. 445, 451, 450 ______
2. 456, 449, 468 ______
3. 470, 462, 468 ______

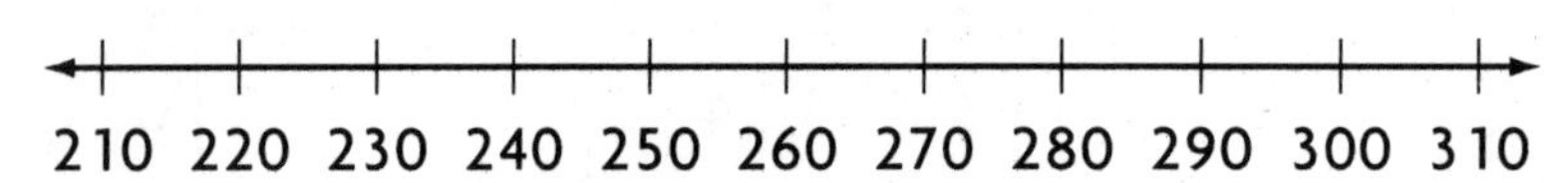

4. 221, 210, 235 ______
5. 305, 275, 255 ______
6. 246, 232, 310 ______

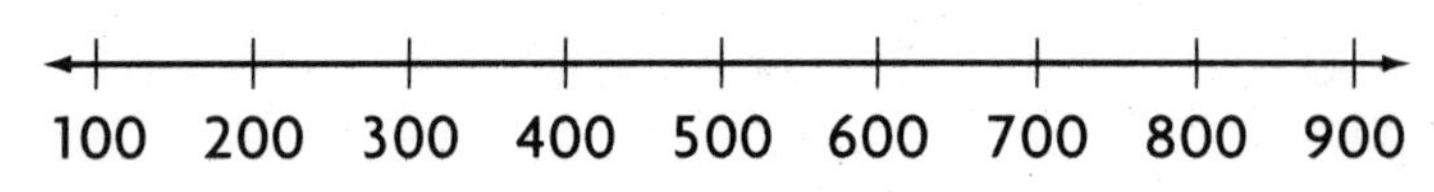

7. 232, 150, 323 ______
8. 560, 595, 499 ______
9. 900, 760, 443 ______

Write the numbers in order from greatest to least.

10. 165, 132, 169 ______
11. 87, 110, 56 ______
12. 254, 124, 304 ______

Mixed Applications

13. Is the letter G the sixth, seventh, or eighth letter of the alphabet? ______

14. Shawn spent $2.89. How much change will he get from $10.00? ______

15. Mona's art teacher visits classrooms whose numbers are 15, 25, 28, 16, 22, 21, 29, and 20. Number the classrooms in order from least to greatest. ______

16. Mindy and her family spent $132, $120, $145, and $125 for food on their four-day vacation. List the amount spent from greatest to least. ______

Name ______________________________

Problem-Solving Strategy

Draw a Picture

Draw a picture to solve.

1. Three girls drive to visit their grandmothers. Sarah drives 125 miles; Tara drives 146 miles; and Lisa drives 136 miles. Who drives the farthest?

____

2. Scott paid $234 for a VCR. Janet paid $324 for a VCR. Who paid more for a VCR? What is the difference between the prices?

____

3. Three gas stations are having a price war. Harry's sells a gallon of regular gas for $1.39. Jake's sells it for $1.42, and Shelly's sells it for $1.36. Order the prices from least to most expensive.

____

4. Mrs. Kay's third-grade class is guessing how many peanuts are in a jar. Josh's guess is 42, and Maggie's guess is 37. There are 40 peanuts. Whose guess is closer?

____

Mixed Applications

Solve.

CHOOSE A STRATEGY

- **Draw a Picture**
- **Act It Out**
- **Find a Pattern**
- **Make a Model**

5. In September, Molly could type 12 words per minute. In June, she could type 45 words per minute. What was the difference in her scores from September to June?

____

6. The distance markers on a long-distance running trail are spaced evenly. The first marker reads 3 miles, and the second reads 6 miles. What does the fourth marker read?

____

7. List the number of drinks sold per day from the greatest to the least.

____

Linda's Lemonade Sales	
Friday	36 cups
Saturday	121 cups
Sunday	110 cups
Monday	45 cups

Name ___________________________

LESSON 10.4

Rounding to Tens and Hundreds

Round each number to the nearest hundred. Use the number line to help you.

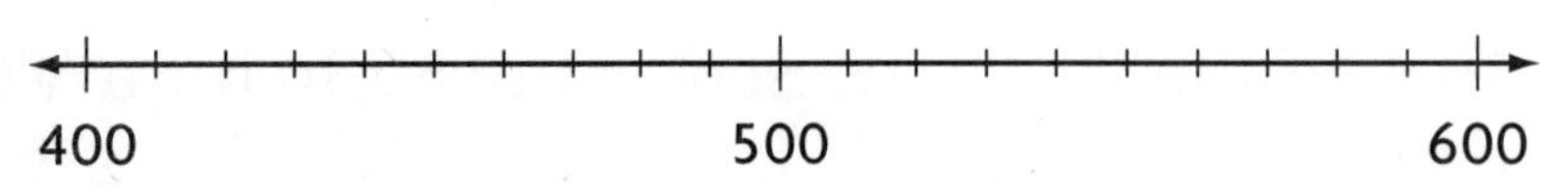

1. 564 ______
2. 412 ______
3. 525 ______
4. 578 ______
5. 445 ______

Round each number to the nearest hundred. You may wish to use a number line.

6. 754 ______
7. 316 ______
8. 283 ______
9. 621 ______
10. 489 ______

Write which two tens or two hundreds each number is between. Then write what the number rounds to.

11. 25 ______ ______ ______
12. 85 ______ ______ ______
13. 250 ______ ______ ______
14. 5 ______ ______ ______

Mixed Applications

15. Tyrone has 24 baseball hats. Does he have closer to 20 hats or 30 hats?

______

16. The movie begins at 4:15. It is two hours long. What time does the movie get out?

______

17. Lindsay has 132 teddy bears. To the nearest hundred, how many teddy bears does Lindsay have?

______

18. Alyssa lives 150 miles from Riverdale, 132 miles from Flagville, and 164 miles from Clayton. Which town is closest?

______

Name ______________________

More About Rounding

Round to the nearest ten or ten dollars.

1. 16 ______
2. 69 ______
3. 44 ______
4. 87 ______
5. 21 ______
6. $25 ______
7. $53 ______
8. $92 ______
9. $66 ______
10. $71 ______

Round to the nearest hundred or hundred dollars.

11. 338 ______
12. 426 ______
13. 845 ______
14. 650 ______
15. 562 ______
16. $135 ______
17. $256 ______
18. $349 ______
19. $750 ______
20. $315 ______

Use the digits 1, 3, and 5. Write a number that rounds to the number given.

21. 300 ______
22. 400 ______
23. 500 ______
24. 100 ______

Mixed Applications

25. Suzy has a total of 259 points on math tests. When rounded to the nearest hundred, do her points round to 200 or 300?
______

26. Maurice has $30.00. Can he buy a book for $16.00, a tape for $12.00, and a magazine for $2.50?
______

27. I am a number between 53 and 73. When rounded to the nearest ten I am 60. What number could I be?
______
______

28. Kelly has 15 compact discs. Monroe has 19 compact discs. Who has more compact discs? What is the difference?
______

Name ______________________________

LESSON 11.1

Making Equal Groups

Use counters to help you find how many in all. Draw a picture of your model.

1. 1 group of 5

2. 2 groups of 6

3. 2 groups of 2

4. 4 groups of 5

5. 8 groups of 2

6. 6 groups of 5

Look at each picture below. Write how many fruits there are in all.

7. 9 groups of 2 = ____________

8. 3 groups of 5 = ____________

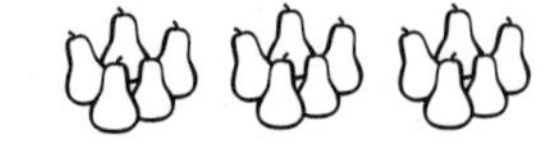

9. 3 groups of 2 = ____________

10. 7 groups of 5 = ____________

11. 5 groups of 2 = ____________

12. 8 groups of 2 = ____________

Mixed Applications

13. Each of 5 children are wearing mittens. How many mittens are there in all?

____

14. One bag holds 5 cookies. How many cookies are in 5 bags?

____

15. Tom needs 15 oranges. Will he have enough oranges if he buys 3 packages with 5 oranges in each package?

____

16. Ann needs 12 pencils. The school store sells pencils in packages of 2. How many packages should she buy?

____

Name ______________________

Multiplying with 2 and 5

Circle the word that best completes each sentence.

1. (Factors, Products) are numbers that you multiply.
2. The answer to a multiplication problem is the (factor, product).

Add and multiply to find how many there are in all.

3. (♠♠♠♠♠) (♠♠♠♠♠) (♠♠♠♠♠)

$5 + 5 + 5 =$ ____

$3 \times 5 =$ ____

4. (⚽⚽) (⚽⚽) (⚽⚽) (⚽⚽) (⚽⚽)

$2 + 2 + 2 + 2 + 2 =$ ____

$5 \times 2 =$ ____

Write the addition sentence and the multiplication sentence for each.

5. (XXXXXXXXX) (XXXXXXXXX)

____

____

6. (JJJJJJ) (JJJJJJ) (JJJJJJ) (JJJJJJ) (JJJJJJ)

____

____

7. (KK) (KK) (KK)

____

____

Find the product. You may wish to draw a picture.

8. $7 \times 5 =$ ____
9. $3 \times 2 =$ ____
10. $8 \times 5 =$ ____
11. $2 \times 2 =$ ____
12. $9 \times 5 =$ ____
13. $2 \times 5 =$ ____
14. $5 \times 6 =$ ____
15. $8 \times 2 =$ ____

Mixed Applications

16. Sue bought 4 packs of film. There are 5 rolls of film in each pack. How many rolls of film did she buy?

____

17. Tom has 5 blue pens and 2 red pens. Sanford has 16 green pens. How many more pens does Sanford have than Tom?

____

Name ______________________________

LESSON 11.2

Problem-Solving Strategy

Draw a Picture

Draw a picture to solve.

1. A tile floor is made of 8 rows of tiles with 5 tiles in each row. How many tiles make up the floor?

____

2. There are 9 baseball players at practice. Each player is thrown 5 pitches. How many pitches are thrown in all?

____

3. Randall has 6 coins in his pocket. The coins are equal to $0.62. What coins are in his pocket?

____

____

4. Lisa watched 9 movies in April, 8 in May, and 5 in June. How many movies did she watch in the three months?

____

Mixed Applications

Solve.

CHOOSE A STRATEGY

- **Act It Out**
- **Make a Model**
- **Find a Pattern**
- **Write a Number Sentence**
- **Draw a Picture**

5. Alice needed 5 ribbons for each costume she made. She made 7 costumes. How many ribbons did Alice need?

____

6. Chris bought a tape that cost $3.95. He gave the clerk $5.00. How much change did he receive?

____

7. Karen used this pattern to make a bead bracelet: 1 red bead and then 4 white beads. She used 30 beads. How many red beads did she use?

____

8. There are 13 people on a train. When the train stops, 6 people get off and 9 people get on. How many people are on the train now?

____

Name ______________________________

Multiplying with 3

Complete the multiplication sentence for each number line.

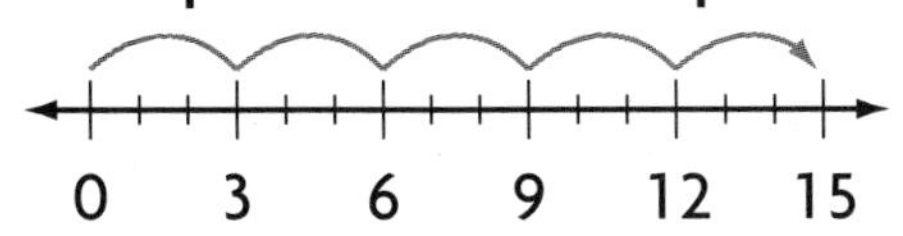

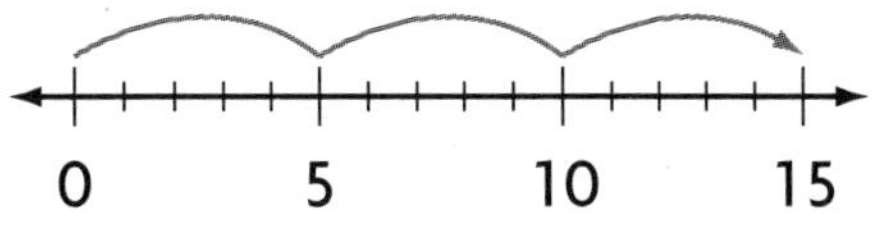

1. $5 \times 3 =$ ____ **2.** $3 \times 5 =$ ____

Use the number line. Find the product.

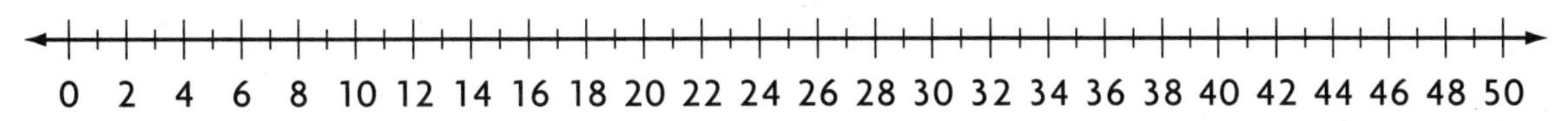

3. $5 \times 5 =$ ____ **4.** $4 \times 3 =$ ____ **5.** $9 \times 3 =$ ____ **6.** $2 \times 3 =$ ____

7. $4 \times 5 =$ ____ **8.** $3 \times 8 =$ ____ **9.** $7 \times 2 =$ ____ **10.** $3 \times 3 =$ ____

11. $9 \times 5 =$ ____ **12.** $6 \times 3 =$ ____ **13.** $2 \times 2 =$ ____ **14.** $5 \times 3 =$ ____

15. $8 \times 2 =$ ____ **16.** $5 \times 9 =$ ____ **17.** $2 \times 9 =$ ____ **18.** $6 \times 5 =$ ____

19. $5 \times 4 =$ ____ **20.** $3 \times 9 =$ ____ **21.** $5 \times 2 =$ ____ **22.** $7 \times 3 =$ ____

Mixed Applications

23. Kia has 6 bags of oranges. There are 3 oranges in each bag. How many oranges does Kia have in all?

____

24. Each of Zack's 3 younger sisters has a tricycle. How many tricycle wheels are there in all?

____

25. Brendon ran 2 miles each day for 8 days. Theo ran 8 miles each day for 2 days. Brendon says he ran more miles in all than Theo did. Is Brendon right?

____

26. Keisha left home at 3:15 and drove for 45 minutes to Diana's house. Then the two friends visited for 2 hours before Keisha left. At what time did Keisha leave?

____

Name ______________________

Multiplying with 1 and 0

Complete the multiplication sentence to show the number of sneakers.

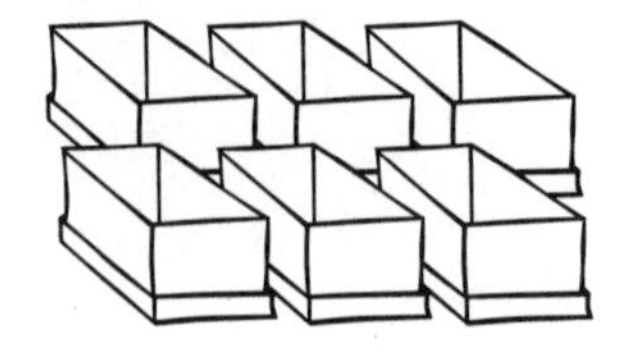

1. $3 \times 1 =$ ____ **2.** $6 \times 0 =$ ____ **3.** $1 \times 2 =$ ____

Find the product.

4. $8 \times 0 =$ ____ **5.** $1 \times 6 =$ ____ **6.** $0 \times 5 =$ ____ **7.** $9 \times 1 =$ ____

8. $1 \times 4 =$ ____ **9.** $0 \times 3 =$ ____ **10.** $1 \times 8 =$ ____ **11.** $0 \times 1 =$ ____

12. $0 \times 0 =$ ____ **13.** $5 \times 1 =$ ____ **14.** $7 \times 0 =$ ____ **15.** $2 \times 5 =$ ____

16. $5 \times 4 =$ ____ **17.** $6 \times 3 =$ ____ **18.** $3 \times 7 =$ ____ **19.** $8 \times 2 =$ ____

Mixed Applications

20. Maya eats 3 pieces of fruit each day. How many pieces of fruit does she eat in a week?

____

21. Rick has 7 nickels. Each nickel is 5 cents. How much are they worth in all?

____

22. Beth spent $4.50 for lunch. Peter spent $1.85 less than Beth did. How much did Peter spend for lunch?

____

23. Mrs. Ramirez reads her son 1 book every day. How many books does she read to him in 9 days?

____

24. Al's lunch cost $2.89. He gave the clerk $3.00. How much change did he receive?

____

25. Jon has 9 coins in his pocket. The coins equal $0.80. What coins are in Jon's pocket?

____

____

Name ____________________

Multiplying with 4

Find the product. You may wish to use a multiplication table.

1. 4×4	2. 1×4	3. 4×7	4. 9×4	5. 4×3	6. 2×4	7. 4×8
8. 0×4	9. 5×4	10. 3×2	11. 2×4	12. 1×4	13. 7×3	14. 9×2
15. 8×2	16. 3×5	17. 5×1	18. 6×5	19. 0×3	20. 1×2	21. 7×0

22. $4 \times 6 =$ ____ 23. $1 \times 0 =$ ____ 24. $5 \times 3 =$ ____ 25. $0 \times 9 =$ ____

26. $4 \times 0 =$ ____ 27. $5 \times 4 =$ ____ 28. $1 \times 0 =$ ____ 29. $8 \times 3 =$ ____

Mixed Applications

Use the table for Problems 30–33.

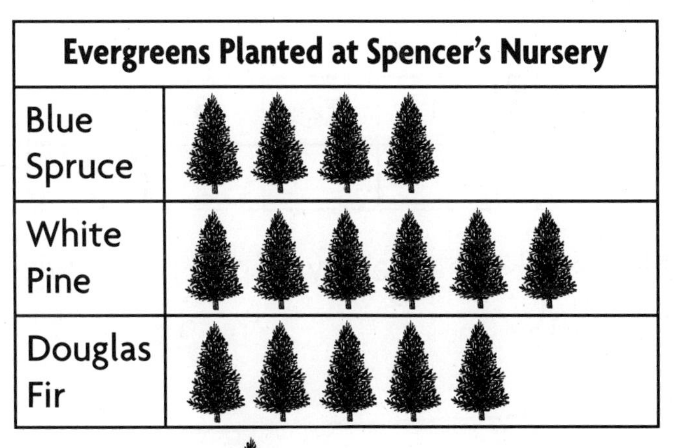

Evergreens Planted at Spencer's Nursery	
Blue Spruce	4 trees
White Pine	6 trees
Douglas Fir	5 trees

Key: Each  equals 5 evergreens.

30. How can you use multiplication to find the total number of each evergreen that the nursery planted?

____

____

31. How many evergreens were planted in all?

____

32. How many more White Pine did the nursery plant than Blue Spruce?

____

33. Which kind of evergreen did Spencer's Nursery plant the most of? How many did they plant?

____

Name ______________________________

LESSON 12.1

Modeling Multiplication 0–6

Vocabulary

Complete.

1. An ____________ shows objects in rows and columns. In arrays for multiplication, the first factor is the number of rows, and the second factor is the number of columns.

Use tiles to make arrays. Find the product.

2. $5 \times 3 =$ ______ 3. $2 \times 6 =$ ______ 4. $4 \times 5 =$ ______

Write the multiplication fact that is shown by each array.

5. ____________ 6. ____________ 7. ____________

Complete the table.

8.

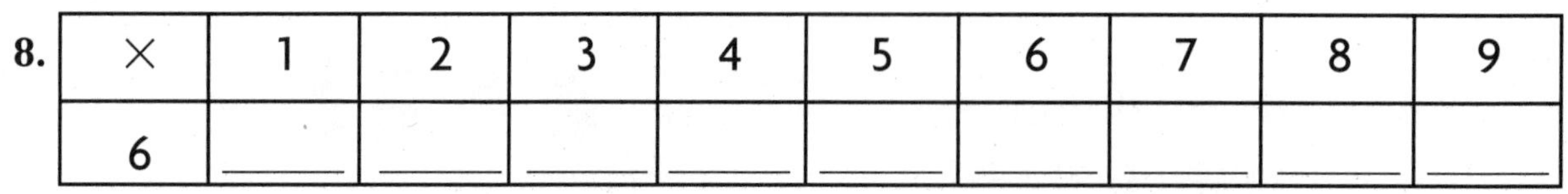

×	1	2	3	4	5	6	7	8	9
6	____	____	____	____	____	____	____	____	____

Name all the arrays you can make with each set of tiles.

9. 8 tiles 10. 9 tiles 11. 16 tiles

Mixed Applications

12. Al bought 4 rows of stamps, with 5 stamps in each row. How many stamps did he buy?

____

13. Jo had 100 stamps. He used 12 of them. How many are left?

____

Name ______________________________

LESSON 12.2

Multiplying with 7

Make two smaller arrays to find each product.

1. 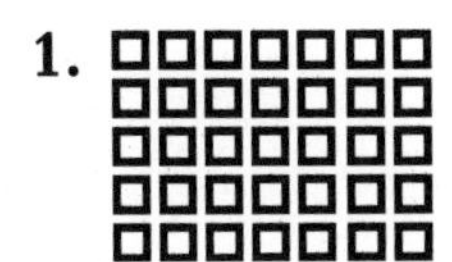

____

____

2. 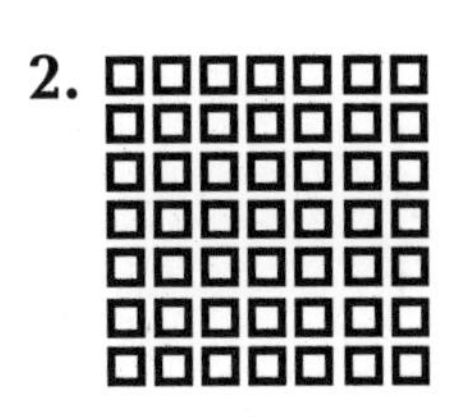

____

____

3.

____

____

Complete the table.

4.

×	1	2	3	4	5	6	7	8	9
7	____	____	____	____	____	____	____	____	____

Find the product.

5. $4 \times 8 =$ ______

6. $7 \times 6 =$ ______

7. $5 \times 9 =$ ______

Mixed Applications

8. Justin is going to camp for 3 weeks. How many days will he spend at camp? How many days will he spend if he leaves camp a week early?

____

9. 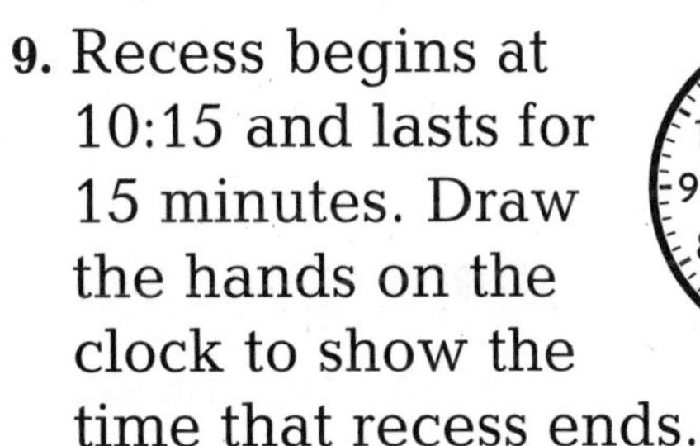

Recess begins at 10:15 and lasts for 15 minutes. Draw the hands on the clock to show the time that recess ends.

10. Patrick bought 5 pencils that cost $0.07 each. How much change should he receive from $1.00?

____

11. The date on Carrie's calendar is April 5. Her birthday will be in 2 weeks. On what date is Carrie's birthday?

____

Name ______________________________

Multiplying with 8

Write two smaller arrays for each array. Find the product.

1.

2. 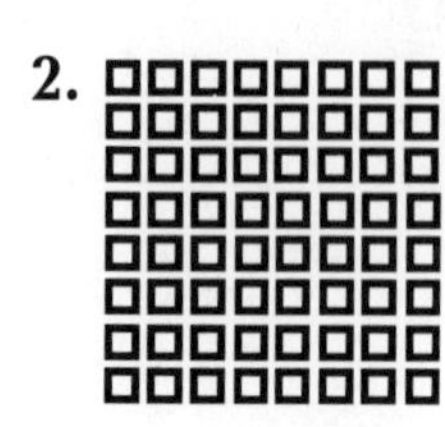

3.

______________ ______________ ______________

______________ ______________ ______________

Complete the table.

4.

×	1	2	3	4	5	6	7	8	9
8	____	____	____	____	____	____	____	____	____

Find the product.

5. $3 \times 6 =$ ____ 6. $8 \times 7 =$ ____ 7. $9 \times 3 =$ ____ 8. $4 \times 7 =$ ____

9. 7×2 10. 6×5 11. 8×7 12. 9×3 13. 4×6

Mixed Applications

For Problems 14–15, use the picture.

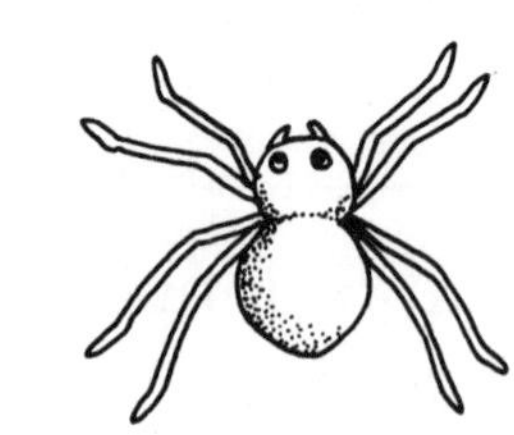

14. How many legs do 3 spiders have in all? 3 ants?

____

15. How many ants in all does it take to have as many legs as 3 spiders? 6 spiders?

____

Name ______________________

Multiplying with 9

Complete the table.

1.

×	1	2	3	4	5	6	7	8	9
9	___	___	___	___	___	___	___	___	___

Find the product.

2. 9×3
3. 4×8
4. 9×8
5. 9×5
6. 6×8
7. 5×7
8. 4×9
9. 8×8
10. 6×9
11. 9×2
12. 4×4
13. 7×7
14. 7×9
15. 2×6
16. 4×6

17. $3 \times 8 =$ ______
18. $5 \times 4 =$ ______
19. $5 \times 1 =$ ______
20. $7 \times 4 =$ ______
21. $6 \times 6 =$ ______
22. $5 \times 9 =$ ______
23. $8 \times 7 =$ ______
24. $3 \times 2 =$ ______

Mixed Applications

For Problems 25–26, use the table.

School Store Price List	
Pencil	5¢
Eraser	8¢
Pen	9¢

25. Ben bought 4 pens at the school store. How much money did he spend?

____

26. Joanna paid for 3 erasers using 6 coins. List the coins that she used.

____

27. What is the difference between the products of 5×10 and 5×9?

____

Name ______________________

LESSON 12.4

Problem-Solving Strategy

Make a Model

Make a model and solve.

1. Jeff puts 5 stamps in each row on an album page. He fills 7 rows. How many stamps in all are there?

____

2. Mrs. Acuna has 3 rows of cups in her cupboard. There are 7 cups in each row. How many cups are in the cupboard?

____

3. Nick cut a pan of brownies so that there are 6 rows, with 8 brownies in each row. How many brownies in all are there?

____

4. Lisa made 4 rows of candles on her father's birthday cake. There are 8 candles in each row. How many candles are on the cake?

____

Mixed Applications

Solve.

CHOOSE A STRATEGY

- **Work Backward**
- **Guess and Check**
- **Act It Out**
- **Make a Model**

5. It is now 10:30. Jack just finished playing baseball for 1 hour. Before that, he read for 30 minutes. At what time did Jack begin reading?

____

6. Mr. McDonald has 20 cows and horses. He has 4 more cows than horses. How many cows does Mr. McDonald have? How many horses?

____

7. Gina's book has 145 pages. Anne's book has 200 pages. How many more pages are there in Anne's book than in Gina's book?

____

8. Dave has 7 coins. He only has pennies, nickels, and dimes. He has more nickels than pennies and more dimes than nickels. How much money does he have?

____

Name ____________________

Completing the Multiplication Table

Use the multiplication table to find the products.

1. 5×7
2. 6×7
3. 8×4
4. 9×3
5. 0×7
6. 6×8
7. 4×7
8. 5×9
9. 4×4
10. 8×3

For Exercises 11–13, use the multiplication table.

11. Circle all of the products in the multiplication table that are odd.

12. Are there more even products or odd products in the table?

____

13. Is the product of two odd factors even or odd?

____

×	0	1	2	3	4	5	6	7	8	9
0	0	0	0	0	0	0	0	0	0	0
1	0	1	2	3	4	5	6	7	8	9
2	0	2	4	6	8	10	12	14	16	18
3	0	3	6	9	12	15	18	21	24	27
4	0	4	8	12	16	20	24	28	32	36
5	0	5	10	15	20	25	30	35	40	45
6	0	6	12	18	24	30	36	42	48	54
7	0	7	14	21	28	35	42	49	56	63
8	0	8	16	24	32	40	48	56	64	72
9	0	9	18	27	36	45	54	63	72	81

Mixed Applications

14. There are 4 rows of desks in Jon's classroom. There are 6 desks in each row. How many desks are there in all?

____

15. Scott has 100 dimes. How many dollars is that equivalent to?

____

16. Stan's class has 29 students, Leo's has 28, and Ben's has 32. Put the number of students in order from least to most.

____

17. Tamala has 2 quarters and 4 dimes. Portia has 3 quarters and 5 nickels. Who has more money?

____

Name ______________________________

Exploring Division

1. Use 15 counters. Put them into as many equal groups as you can. Write division sentences to record what you did.

____

____

Complete the table. Use counters to help you.

	Counters	How many groups?	How many in each group?
2.	10	2	
3.	12		6
4.	16	4	
5.	18		6
6.	21	3	

Mixed Applications

7. Jack spent 18¢ for pencils. How many pencils did he buy?

____

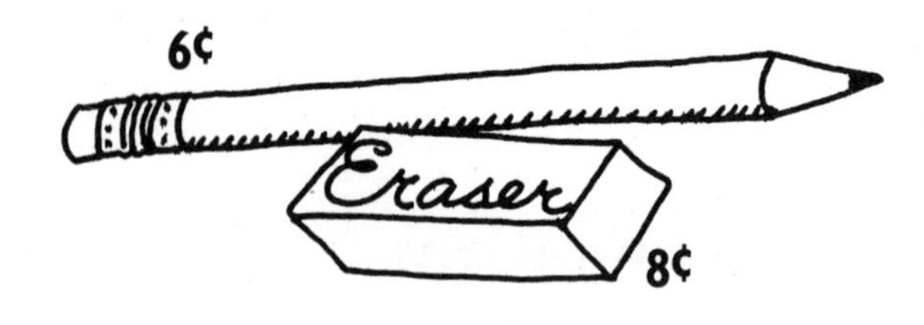

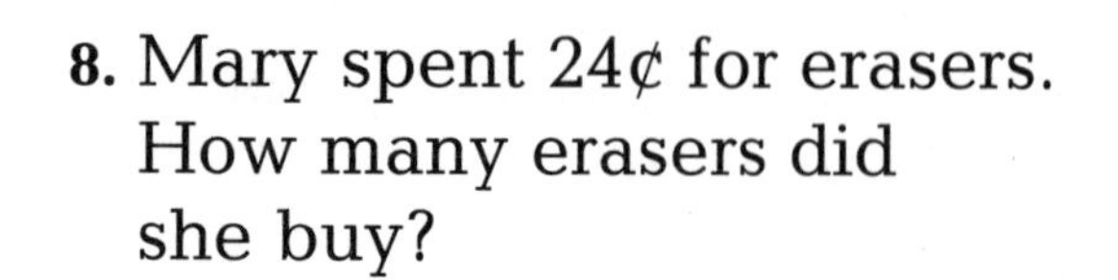

8. Mary spent 24¢ for erasers. How many erasers did she buy?

____

9. Bill bought 6 pencils. How much change should he receive from a $1 bill?

____

10. There are 24 students who are working in groups of 4 on science projects. How many groups are there?

____

11. The product of two numbers is 24. Their difference is 5. What are the two numbers?

____

Name ____________________

Connecting Subtraction and Division

Show how you can use subtraction to solve.

1. $12 \div 3 =$ ____

2. $20 \div 4 =$ ____

3. $30 \div 5 =$ ____

4. $6 \div 2 =$ ____

Write the division sentence shown by the repeated subtraction.

5. $18 - 3 = 15$, $15 - 3 = 12$, $12 - 3 = 9$, $9 - 3 = 6$, $6 - 3 = 3$, $3 - 3 = 0$

6. $18 - 6 = 12$, $12 - 6 = 6$, $6 - 6 = 0$

7. $10 - 2 = 8$, $8 - 2 = 6$, $6 - 2 = 4$, $4 - 2 = 2$, $2 - 2 = 0$

8. $24 - 4 = 20$, $20 - 4 = 16$, $16 - 4 = 12$, $12 - 4 = 8$, $8 - 4 = 4$, $4 - 4 = 0$

Mixed Applications

9. Cathy uses 18 stickers to make cards. She puts 3 stickers on each card. How many cards can she make in all?

10. Paula, Anita, Alan, and Jesse shared a bag of 36 marbles equally. How many marbles did each person get?

Name ______________________________

LESSON 13.3

Relating Multiplication and Division

Vocabulary

Underline the correct word to complete each sentence about division.

1. The *dividend/quotient* is the number being divided.
2. The *quotient/divisor* is the number that divides the dividend.
3. The answer is the *divisor/quotient.*
4. Multiplication and division are *identical/inverse*, or opposite, operations.

Use the array to find the quotient.

5. 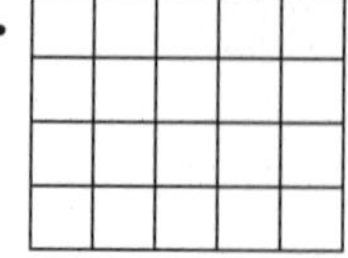

$20 \div 4 =$ ____

6. 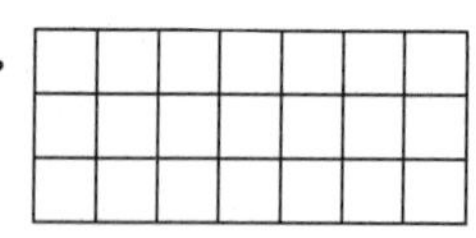

$21 \div 3 =$ ____

7. 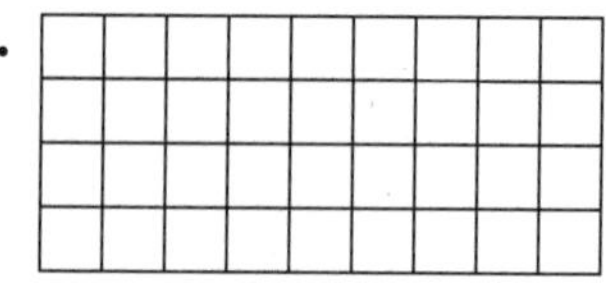

$36 \div 4 =$ ____

Write the missing factor for each number sentence.

8. $3 \times$ ____ $= 27$ $27 \div 3 =$ ____
9. $5 \times$ ____ $= 40$ $40 \div 5 =$ ____
10. $6 \times$ ____ $= 18$ $18 \div 6 =$ ____
11. $8 \times$ ____ $= 32$ $32 \div 8 =$ ____

Check each division with multiplication. Show your work.

12. $30 \div 6 =$ ____

____

13. $18 \div 2 =$ ____

____

14. $25 \div 5 =$ ____

____

Mixed Applications

15. Tennis balls are sold in sets of 3. How many sets must Mr. Campbell buy in order to have 15 tennis balls?

____

16. Tom has 3 dimes, 8 nickels, and 3 pennies. How much more money does he need to buy a 75¢ ice cream cone?

____

Name ______________________

LESSON 13.4

Fact Families

Vocabulary

Fill in the blank to complete the sentence.

1. A ______________ is a set of related multiplication and division sentences that use the same numbers.

Write the other three sentences that belong in the fact family.

2. $6 \times 3 = 18$

3. $4 \times 5 = 20$

4. $2 \times 7 = 14$

Write the fact family for each set of numbers.

5. 4, 9, 36

6. 8, 3, 24

7. 6, 4, 24

8. 6, 6, 36

9. 7, 7, 49

10. 5, 5, 25

Mixed Applications

11. Beth cuts a 36-inch length of ribbon into 4 equal pieces. How long is each piece?

12. Frank spent $2.75. He then had $2.50 left. How much money did he have to begin with?

Name ______________________________

LESSON 13.5

Practicing Division Facts Through 5

Write the multiplication fact you can use to find each quotient. Write the quotient.

1. $30 \div 6 =$ ____ ________________
2. $25 \div 5 =$ ____ ________________
3. $27 \div 3 =$ ____ ________________
4. $21 \div 7 =$ ____ ________________
5. $16 \div 4 =$ ____ ________________
6. $36 \div 9 =$ ____ ________________
7. $18 \div 3 =$ ____ ________________
8. $45 \div 9 =$ ____ ________________
9. $28 \div 4 =$ ____ ________________
10. $12 \div 4 =$ ____ ________________
11. $27 \div 9 =$ ____ ________________
12. $20 \div 5 =$ ____ ________________
13. $32 \div 8 =$ ____ ________________
14. $40 \div 5 =$ ____ ________________
15. $21 \div 3 =$ ____ ________________
16. $15 \div 5 =$ ____ ________________
17. $9 \div 3 =$ ____ ________________
18. $16 \div 2 =$ ____ ________________

Mixed Applications

19. Some crayons are divided equally among 4 students. Each student gets 9 crayons. How many crayons are there in all?

    ____

20. A music class began at 10:45. The class spent 20 minutes singing and 10 minutes listening to music. At what time did the class end?

    ____

21. A store sells 4 balloons for $0.45. How much would 8 balloons cost?

    ____

22. Phil wants to trade in his 40 nickels for quarters. How many quarters can he get?

    ____

Name ______________________________

Choosing Division or Multiplication

Circle ***a*** or ***b*** to show which number sentence you would use to solve each problem. Then write the answer.

1. There are 9 tables in the art room. There are 3 students sitting at each table. How many students are there in all?

   a. $9 \times 3 = 27$

   b. $9 \div 3 = 3$

   ____

2. Kyla has 10 shoes in her closet. How many pairs of shoes does she have?

   a. $10 \times 2 = 20$

   b. $10 \div 2 = 5$

   ____

3. Dave wants to buy 8 batteries. The batteries are sold in packages of 4. How many packages of batteries should Dave buy?

   a. $8 \times 4 = 32$

   b. $8 \div 4 = 2$

   ____

4. Pam went to an art class on 8 Saturdays. Each class was 2 hours long. How many hours did Pam spend in the art classes in all?

   a. $8 \times 2 = 16$

   b. $8 \div 2 = 4$

   ____

Mixed Applications

5. Diana bought the ball and jump rope. She paid with a $1 bill. What is the least number of coins she could receive in change?

   ____

   ____

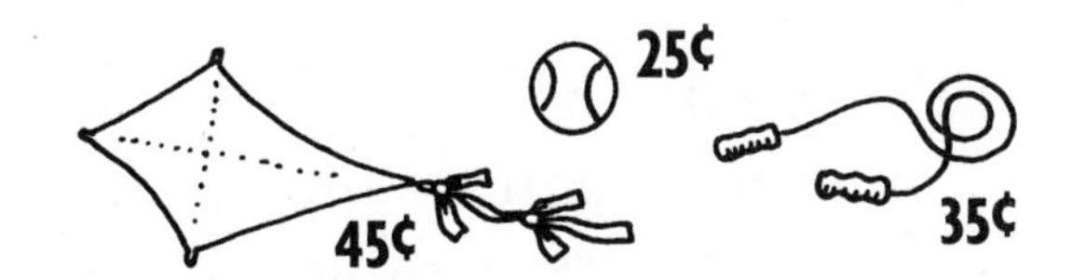

6. Josh paid for the kite, using all nickels. How many nickels did he spend?

   ____

7. Diana practiced jumping rope every day for 3 weeks. How many days did she practice?

   ____

Name ______________________

LESSON 13.6

Problem-Solving Strategy

Write a Number Sentence

Write a number sentence to solve. Then write the answer.

1. Mrs. Scott bought 3 packages of hot dogs. Each package has 8 hot dogs. How many hot dogs did she buy in all?

____

2. A class of 27 students is working in groups of 3 on an art project. How many groups are there?

____

3. Melissa took 24 photographs. She put 4 photographs on each page of her album. How many pages did she use?

____

4. Tim planted 5 rows of corn. There are 6 corn plants in each row. How many corn plants are there in all?

____

Mixed Applications

Solve.

CHOOSE A STRATEGY

- Guess and Check
- Act It Out
- Make a Model
- Draw a Picture
- Write a Number Sentence

5. How many more bicycles were sold in May than in April?

____

Bicycle Sales	
March	78
April	125
May	209

6. There are 5 chairs on each side of a large square table. How many chairs are there in all?

____

7. How many wheels does Mr. Jackson need to build 6 wagons?

____

8. Robert has 24 car and airplane models. He has 2 times as many car models as airplane models. How many car models does he have?

____

9. A music room has 4 rows of chairs. Each row has 8 chairs. If 28 students sit in the music room, how many chairs will not be used?

____

Name ____________________

Modeling Division Using Arrays

For Exercises 1–6, use square tiles. Write the division sentence that records what you did.

1. How many groups of 5 are in 20?

 ____

2. How many groups of 4 are in 16?

 ____

3. How many groups of 6 are in 18?

 ____

4. How many groups of 3 are in 12?

 ____

5. Use 28 tiles. Make an array that has 7 rows.

 ____

6. Use 32 tiles. Make an array that has 4 rows.

 ____

Mixed Applications

7. In Sandy's classroom, there are 4 rows of desks and 6 desks in each row. How many desks are there?

 ____

8. Peter arranges his 18 model cars in 3 rows. How many cars are there in each row?

 ____

9. Lauren has $4.30 in her wallet. Ryan has $3.98. How much more money does Lauren have than Ryan?

 ____

10. A class of 22 students goes outside to play softball. If 4 children decide not to play, how many teams of 9 can the class form?

 ____

11. Louise left her house at 7:45. It took her 30 minutes to walk to school. Draw the hands on the clock to show what time Louise arrived at school.

Name ______________________

LESSON 14.2

Dividing Using 0 and 1

Find the quotient.

1. $7 \div 7 =$ ____
2. $0 \div 5 =$ ____
3. $4 \div 1 =$ ____
4. $8 \div 1 =$ ____
5. $6 \div 6 =$ ____
6. $0 \div 3 =$ ____
7. $2 \div 2 =$ ____
8. $0 \div 8 =$ ____
9. $2 \div 1 =$ ____
10. $0 \div 4 =$ ____
11. $3 \div 1 =$ ____
12. $5 \div 5 =$ ____
13. $4 \div 4 =$ ____
14. $9 \div 1 =$ ____
15. $0 \div 2 =$ ____
16. $7 \div 1 =$ ____
17. $9 \div 9 =$ ____
18. $6 \div 1 =$ ____
19. $0 \div 1 =$ ____
20. $0 \div 9 =$ ____
21. $3 \div 3 =$ ____

Mixed Applications

For Problems 22–26, use the table.

Item	Number in a Package	Cost Per Package
Tennis ball	3	$2.50
Marble	24	$1.00
Bouncy ball	4	$1.50
Baseball	1	$2.98

22. Rachel bought 4 cans of tennis balls. How many tennis balls did she buy?

____

23. Doug bought 6 tennis balls and 8 bouncy balls. How much money did he spend?

____

24. Justin bought a baseball and a bag of marbles. What was his change from a $5 bill?

____

25. Grace bought a bag of marbles. She kept 20 of the marbles for herself. She divided the rest of the marbles equally among 4 friends. How many marbles did each friend get?

____

26. Mrs. Kelsey bought 2 packages of bouncy balls. She divided the balls equally among 8 children. How many bouncy balls did each child get?

____

Name ____________________

LESSON 14.3

Using the Multiplication Table to Divide

Use the multiplication table to find each quotient.

1. 15 ÷ 3 = ____	2. 21 ÷ 7 = ____	3. 72 ÷ 9 = ____
4. 81 ÷ 9 = ____	5. 27 ÷ 3 = ____	6. 18 ÷ 6 = ____
7. 48 ÷ 8 = ____	8. 24 ÷ 8 = ____	9. 42 ÷ 7 = ____
10. 35 ÷ 5 = ____	11. 14 ÷ 2 = ____	12. 30 ÷ 6 = ____
13. 40 ÷ 8 = ____	14. 36 ÷ 9 = ____	15. 18 ÷ 9 = ____

Mixed Applications

For Problems 16–20, use the pictures.

16. The students in Mr. Rice's class can earn tickets and trade them in for prizes. Jack has 8 tickets. How many bookmarks can he get?

____

4 tickets

7 tickets

9 tickets

17. Barb has 21 tickets. How many pencils can she get?

____

18. How many tickets must Judy use to get 3 notebooks?

____

19. Mark traded in 25 tickets for 5 prizes. What prizes did he get?

____

20. How many tickets would a student need to earn to get 2 of each prize?

____

21. The clock shows the time John started a walk. He walked for 1 hour and 15 minutes. At what time did John finish his walk?

____

Name ____________________

LESSON 14.4

Practicing Division Facts Through 9

Complete the tables.

1.

÷	24	36	12	28	16
4					

2.

÷	35	45	40	25	15
5					

3.

÷	24	27	18	3	15
3					

4.

÷	48	12	24	36	18
6					

Find the quotient.

5. $36 \div 6 =$ ____
6. $24 \div 8 =$ ____
7. $42 \div 7 =$ ____
8. $56 \div 8 =$ ____
9. $63 \div 7 =$ ____
10. $14 \div 2 =$ ____
11. $8 \div 8 =$ ____
12. $48 \div 6 =$ ____
13. $72 \div 9 =$ ____
14. $4 \div 1 =$ ____
15. $45 \div 5 =$ ____
16. $21 \div 3 =$ ____

Mixed Applications

For Problems 17–19, use the pizza-price list.

JILL'S PIZZA SHOP		
Size	Number of Slices	Price
Small	4	$2.75
Medium	6	$4.95
Large	8	$8.25

17. A group of 9 friends wants to buy enough pizza for each person to have 2 slices. How many medium pizzas should they buy?

____

18. Mr. Welch bought 2 small pizzas and 1 large pizza. What was his change from a $20 bill?

____

19. Mrs. Mason bought 3 large pizzas for a party. At the end of the party, 5 slices of pizza were left. How many slices of pizza had been eaten?

____

Name Breanity Acevedo December 2 2010

LESSON 14.4

Problem-Solving Strategy

Make a Table

Make a table to solve each problem.

1. Jason earns $3.00 an hour for raking leaves. Jason earned $12.00 for raking leaves in September, $27.00 in October, and $15.00 in November. How many hours did he work each month?

____

2. Mr. Stone makes toy cars. He has a package of 36 wheels for large cars, 28 wheels for medium cars, and 16 wheels for small cars. How many of each size car can he make?

____

____

3. It takes Gina 6 minutes to make a greeting card. She made cards for 30 minutes in the morning, 12 minutes in the afternoon, and 18 minutes in the evening. How many cards did she make during each time?

____

4. The Colorado River is 1,450 miles long. The Columbia River is 1,243 miles, the Mississippi River is 2,340 miles, and the Ohio River is 981 miles. List the rivers in order from shortest to longest.

____

____

Mixed Applications

Solve.

CHOOSE A STRATEGY

- Guess and Check
- Act It Out
- Make a Model
- Make a Table
- Write a Number Sentence

5. Stan had a length of rope. He cut the rope in half, and then he cut each piece in half again. Each piece of rope is 7 inches long. How long was Stan's rope before he cut it?

____

6. Don has a pocketful of quarters, dimes, and nickels. He has more quarters than dimes and more dimes than nickels. He has $1.00. What coins does he have?

____

Name ______________________________

Choosing the Operation

Circle *a* or *b* to show which number sentence you would use to solve each problem.

1. There are 9 mice in each cage. There are 3 cages. How many mice are there in all?

   a. $9 \times 3 = 27$

   b. $9 \div 3 = 3$

2. Izzy and Tom are cats. Izzy weighs 9 pounds and Tom weighs 12 pounds. How much more does Tom weigh than Izzy?

   a. $12 + 9 = 21$

   b. $12 - 9 = 3$

3. Mrs. Ellis buys 9 cans of cat food. She already has 8 cans of cat food at home. How many cans does she have now?

   a. $9 \times 8 = 72$

   b. $9 + 8 = 17$

4. Mr. Davis has 24 goldfish. He puts 8 fish in each fish bowl. How many fish bowls does he use?

   a. $24 \div 8 = 3$

   b. $24 - 8 = 16$

Mixed Applications

For Problems 5–7, use the pictures.

Oranges
3 for $1.25

Apples
6 for $1.59

5. Mrs. Spencer wants to buy 24 apples. How many bags of apples should she buy?

   ____

6. Mr. Long bought 1 bag of apples and 1 bag of oranges. He paid with a $10 bill. How much change should he receive?

   ____

7. Mrs. Szabo is making a large fruit salad. The recipe calls for 2 peaches for every 3 oranges. How many bags of oranges will she need if she uses 10 peaches in the salad? (You may *make a table* to solve.)

   ____

Name ______________________

Collecting and Organizing Data

Vocabulary

Write the correct letter from Column 2.

_____ 1. A table that uses tally marks to show how often something happens.

_____ 2. Information about people or things.

_____ 3. A table that uses numbers to show how often something happens.

a. data

b. frequency table

c. tally table

Make a tally table. List four types of pets in the table. Then ask each classmate which pet he or she likes best. Make a tally mark beside the name of the pet. Then make a frequency table of the same data. Answer Problems 4–5.

4. Which type of pet did the most classmates choose? Which did the fewest choose?

____

5. Compare your tables with those of your classmates. Did everyone get the same results?

____

Mixed Applications

For Problems 6–7, use the table.

6. Change this tally table into a frequency table.

7. Which sport did the most students choose? the fewest choose?

____

Favorite Sports	
Baseball	卌 /
Skating	卌 卌 ///
Soccer	卌 卌 /
Tennis	卌 ///

Name ____________________

LESSON 15.2

Recording Data

Vocabulary

Complete the sentence.

1. An ____________________ is a test done in order to find out something.

For Problems 2–5, tell what kind of table should be used. Write *tally table* or *frequency table.*

2. Sonia did an experiment with a coin. Now she wants to show what happened.

____

3. Sam is getting ready to do a number cube experiment. He will need to record the number that is rolled each time.

____

4. Cindy just finished an experiment. She rolled a number cube twice. She wants to show the class how the experiment turned out.

____

5. Mr. James is about to show his class how to do an experiment with a coin and a spinner. He wants to show how to record what happens.

____

Mixed Applications

6. Tammy did an experiment with a number cube. She rolled it 20 times. She rolled a 1 four times, a 2 one time, a 3 three times, a 4 five times, a 5 two times, and a 6 five times. Make a tally table to show what happened.

7. Julio buys a truck for $0.65. He gives the clerk $1.00. The clerk does not have any quarters for change. What is the least number of coins Julio can receive?

____

Name ______________________

Problem-Solving Strategy

Make a Table

Make a table to solve.

1. Karen and Jose are doing an experiment with a spinner and a coin. They spin the pointer on the spinner and flip the coin. Then they record the results. They will repeat this experiment 15 times. Show how they could organize a table about their experiment.

2. Phillip is doing an experiment with two coins. In the experiment, he will flip each coin 25 times and record the results after each flip. Show how he could organize a table about his experiment.

Mixed Applications

Solve.

CHOOSE A STRATEGY

- **Write a Number Sentence**
- **Act It Out**
- **Guess and Check**
- **Work Backward**

3. Sid biked 4 miles less than John. Together, they biked 12 miles. How many miles did each bike?

____

4. In line, Pat is ahead of Mary. Tom is behind Sally. Tom is between Sally and Pat. What is the order of the children in line?

____

Name ____________________

LESSON 15.3

Understanding Collected Data

Vocabulary

Write the correct letter from Column 2.

_____ 1. A set of questions that a group of people are asked. **a.** survey

_____ 2. The set of answers from a survey. **b.** results

For Problems 3–6, use the survey results in the tally table.

OUR FAVORITE GAMES	
Game	**Tallies**
Follow-the-Leader	~~||||~~ ||
Jump Rope	~~||||~~ ~~||||~~ ~~||||~~	
Tether Ball	~~||||~~ ~~||||~~	
Four-Square	||||

3. List the games in order from the most favorite to the least favorite.

____

____

____

4. How many people answered the survey?

____

5. How many more people like jump rope than four-square?

____

6. How many more people like jump rope than follow-the-leader?

____

Mixed Applications

7. Each of 5 students gets 6 crayons. How many crayons are there?

____

8. Henry walked 300 yards at lunch. He walked 300 yards twice after school. How many yards did he walk in one day?

____

Name ______________________________

LESSON 15.4

Grouping Data in a Table

For Problems 1–5, use the table.

DOGS OWNED BY STUDENTS				
	Black Hair	White Hair	Brown Hair	Gold Hair
Short Hair	3	4	1	3
Medium Hair	2	2	0	1
Long Hair	1	3	3	2

1. How many dogs have short, brown hair?

____

2. How many dogs have medium hair?

____

3. How many dogs have white hair?

____

4. What color of hair do only 4 dogs have?

____

5. How many dogs are owned by the class?

____

6. Look at the marbles. Make a table to group the marbles.

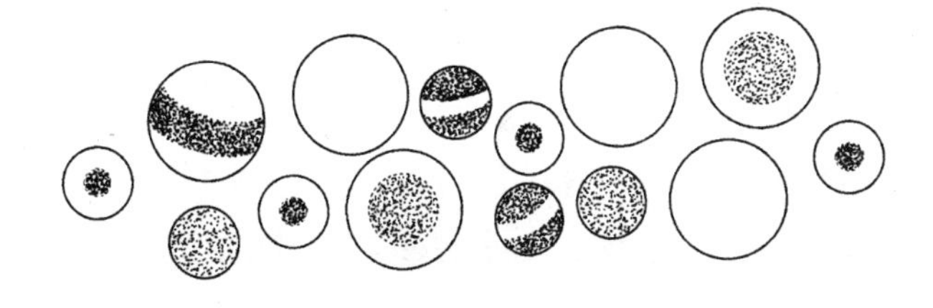

Mixed Applications

7. On Toby's soccer team there are 11 girls and 8 boys. Of the girls, 6 are good defenders and the rest are good forwards. Of the boys, 5 are good defenders and the rest are good forwards. Make a table to group the players on the soccer team.

PEOPLE WITH GLASSES		
	With Glasses	Without Glasses
Boys	5	7
Girls	6	8

8. Write a question about the data in this table.

____

____

Name ____________________

LESSON 16.1

Reading Pictographs

Vocabulary

1. A ____________________ shows data using pictures that stand for more than one thing.

2. The ____________________ at the bottom of the pictograph tells how many each picture stands for.

For Problems 3–6, use the pictograph.

Our Favorite Crayon Colors	
Red	🖍🖍🖍🖍🖍🖍
Blue	🖍🖍🖍
Violet	🖍🖍🖍🖍🖍🖍🖍
Green	🖍
Yellow	🖍🖍🖍🖍

Key: Each 🖍 stands for 2 students.

3. How many students does each crayon picture stand for?

____

4. How many students like each color of crayon?

____

____

5. Which color of crayon do the most students like the best? the fewest students?

____

6. How many more students like yellow crayons best than like green crayons best?

____

Mixed Applications

7. For the class, Mr. Thomas spent $3.56 on crayons and $2.45 on paper. How much did he spend in all on materials for the class?

____

8. The art class started at 2:30. It ended 1 hour and 30 minutes later. At what time did the art class end?

____

Name ______________________________

LESSON 16.2

Making a Pictograph

Think of an idea for making a pictograph. You may want to take a survey or collect data about a subject that interests you.

Collect the data, and then make a pictograph in the space below. Decide on a symbol and key for the graph. Include a title and labels. Show your pictograph to your class and tell about it. Answer problems 1–2 as you tell about your pictograph.

Survey Results

Pictograph

1. Tell how you decided on the subject of your pictograph.

____

____

2. Explain how you chose a symbol and key for your pictograph. Do your classmates agree with your choices?

____

Mixed Applications

3. Fred paid for a new book with a $10.00 bill. His change was $5.37. How much did the book cost?

____

4. In the relay, Sally beat James. Sam beat Mary. Mary beat Sally. In what order did they finish?

____

Name ________________________________

LESSON 16.3

Reading Bar Graphs

Vocabulary

Write *true* or *false*.

__________ 1. *Bar graphs* use lines to stand for data.

__________ 2. A *vertical* bar graph has bars that go up.

__________ 3. A *horizontal* bar graph has bars that go across from left to right.

__________ 4. The *scale* on a bar graph has no use.

For problems 5–8, use the bar graph.

5. What type of bar graph is this?

____

6. How many students named lions as their favorite stuffed animal? frogs? dogs?

____

____

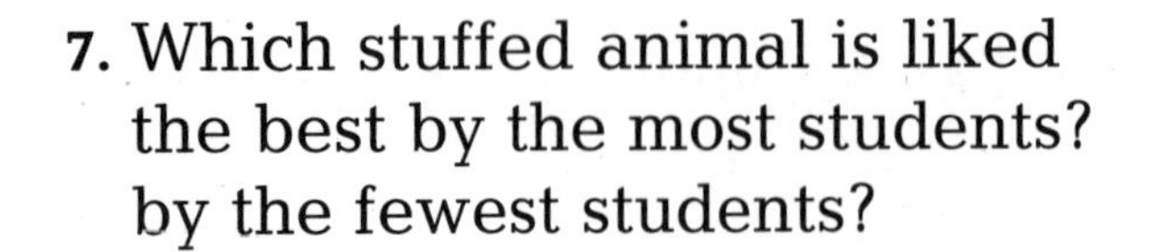

7. Which stuffed animal is liked the best by the most students? by the fewest students?

____

8. How many students named seals as their favorite stuffed animal?

____

Mixed Applications

9. Find the next three numbers in the pattern:
5, 9, 13, 17, 21, 25

____

10. John has done 4 math problems on his homework. Sybil has done 2 times as many math problems. How many math problems has Sybil done?

____

Name ____________________

Making Bar Graphs

Make a bar graph of the data in the table at the right. Use a scale numbered by 2's (0, 2, 4, 6, 8, 10, 12). Remember to title and label the graph.

FAVORITE DRINKS OF MR. HALE'S CLASS	
Water	4
Lemonade	6
Punch	2
Milk	5
Juice	8
Soda	12

Use your graph above to answer Problems 1 and 2.

1. What does the graph show?

____

2. What scale is used in the graph?

____

Mixed Applications

3. Tony's book has 24 pages. Each story in it is 4 pages long. How many stories are in the book?

____

4. Cheryl spent $2.25 for lunch. She got $2.75 in change. How much money did Cheryl have for lunch?

____

Name ______________________________

Comparing Data

For Problems 1–4, use the pictograph.

Coin Collection	
Darryl	◎◎◎◎
Sam	◎◎◎◎◎
Elizabeth	◎◎◎◎◎◎◎◎
Morgan	◎
Dirk	◎◎◎◎◎◎
Michael	◎◎◎◎◎◎◎◎◎◎◎◎

Key: Each ◎ stands for 4 coins.

1. How many coins does Darryl have in his collection? Morgan? Elizabeth?

____

2. Who has the most coins in their collection? the fewest?

____

3. Who has more coins—Sam or Darryl?

____

4. How many more coins does Michael have than Dirk?

____

Mixed Applications

5. At the zoo, there are 5 monkeys in one cage. There are 3 times as many birds in another cage. How many birds are there?

____

6. Sonia arrived 1 hour early for the party. The time was 6:45. At what time did the party begin?

____

7. The Middville Chess Club has 16 members playing chess. Each game needs two people. How many games are being played?

____

8. At Sebastian Music Store, there are 58 violins, 124 guitars, and 245 trombones. How many instruments in all are at the music store?

____

9. Five people are in a line. Harry is first. Ted is in front of Beth. There are two people between Harry and Ted. What is Ted's position in line?

____

Name ______________________________

Problem-Solving Strategy

Use a Graph

For Problems 1–4, use the graphs.

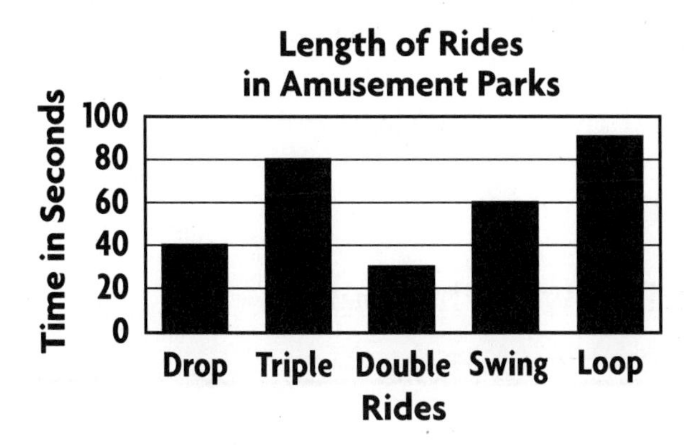

Magazine Subscriptions Sold	
Shirley	
Fred	
Morton	
Amber	
Mack	
Marsha	

Key: Each stands for 2 magazines.

1. Which ride lasts the longest? the shortest?

____

2. How long would you ride if you went on Swing, and twice on Triple?

____

3. How many magazines did Fred sell?

____

4. How many more magazines did Amber sell than Morton?

____

Mixed Applications

Solve.

CHOOSE A STRATEGY

- Find a Pattern
- Guess and Check
- Act It Out
- Use a Graph

5. It was 3:30 when Rolly began his homework. He finished in 10 minutes. Five minutes after he finished he remembered to put the homework in his notebook. At what time did he put the homework in his notebook?

____

6. Sean bought a new reading book for $1.20. He gave the cashier 6 coins. What coins did Sean use?

____

____

____

Name ______________________________

LESSON 17.1

Certain and Impossible

Vocabulary

Fill in the blank with the correct word.

event   certain   impossible

1. An event is ______________ if it will never happen.
2. An ______________ is something that happens.
3. An event is ______________ if it will always happen.

Read each event. Tell whether the event is *certain* or *impossible* to happen.

4. Pencils will fall from the sky.

____

5. Winter in Alaska is cold.

____

6. You will walk to the moon tonight.

____

7. Putting your hand in boiling water will burn you.

____

For Problems 8–9, use the numbered tile. Tell whether each event is *certain* or *impossible*.

1	3	3
1	5	7
3	5	7

8. dropping a coin on an odd number ______________________
9. dropping a coin on a number greater than 9 ______________________

Mixed Applications

10. A block has the numbers 1 through 4. Is it certain or impossible that you will roll a number less than 5?

____

11. Sophie has 5 dimes, 2 nickels,and 3 pennies. How much money does she have?

____

Name ____________________

Recording Possible Outcomes

Vocabulary

Write the correct letter from Column 2.

Column 1		Column 2
1. most likely	____	a. something that has a chance of happening
2. possible outcome	____	b. an event that has a lesser chance of happening compared to other events
3. least likely	____	c. an event that has a greater chance of happening than other events

---

List the possible outcomes of each event.

4. trying on a pair of pants at the store

____

5. picking a number that is even from the numbers 2–12

____

Tell which event is *most likely* to happen.

6. dropping a marker on one of these squares

<table>
<tr><td>1</td><td colspan="2" rowspan="2">11</td></tr>
<tr><td>3</td></tr>
<tr><td>5</td><td>7</td><td>9</td></tr>
</table>

____

7. picking a number from this bag

____

Mixed Applications

8. There is a bag of 4 blue balls, 2 green balls, and 1 red ball. Which color is most likely to be picked?

____

9. There are 30 people in the movie theater. During the movie 6 people leave, but 2 people return. How many people are there at the end of the movie?

____

Name ______________________________

LESSON 17.2

Problem-Solving Strategy

Make a List

Make a list to solve.

1. Eileen has a blue, a red, and a yellow shirt. She has white, black, and gray pants. What are the possible combinations of shirts and pants she might pull from the closet?

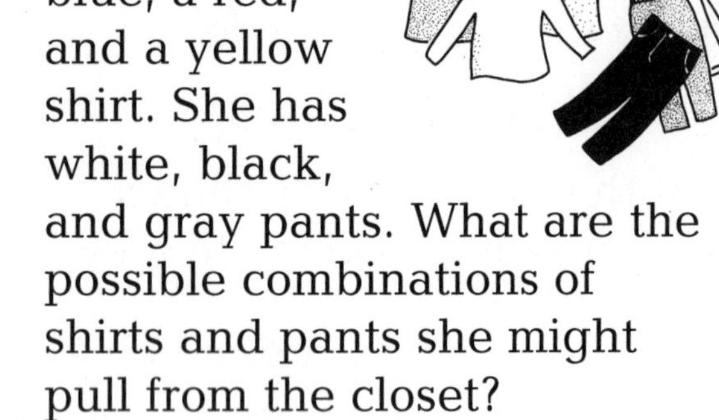

____

____

____

____

2. Tony has two pizzas. They are divided into cheese (C), pepperoni (P), mushroom (M), and onion (O). What are the possible outcomes if he chooses one piece from each pizza at the same time?

____

Mixed Applications

Solve.

CHOOSE A STRATEGY

- **Find a Pattern**
- **Use a Graph**
- **Work Backward**
- **Write a Number Sentence**

3. Find the numbers that come next in the pattern: 168, 166,

   164, ______, 160, ______,

   ______, 154, 152, ______, 148.

4. On Wednesdays, Jana leaves the pool at 1:30. She swims for 50 minutes while she is there. At what time does Jana get to the pool?

____

5. In this ice cream survey, what flavor was bought the most often? least often?

____

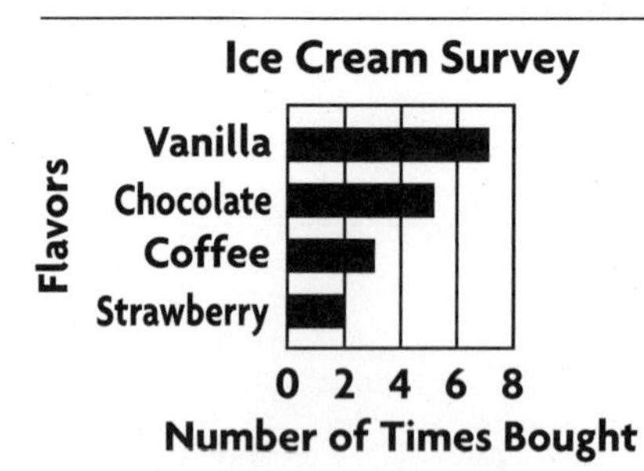

Name ______________________

Recording Results of an Experiment

Read the following experiment. Record the results in the tally table, and answer the questions.

Marsha has a bag filled with 20 balls. There are 7 blue, 2 green, 4 yellow, and 7 red balls. She picks a ball from the bag 10 times to see what color comes out.

Pick #1–red
Pick #2–blue
Pick #3–red
Pick #4–yellow
Pick #5–green
Pick #6–red
Pick #7–blue
Pick #8–yellow
Pick #9–red
Pick #10–blue

MARSHA'S EXPERIMENT

Color	Tally
Red	
Blue	
Yellow	
Green	

1. What color did she pick the most?

____

2. What color did she pick the least?

____

3. Why do you think this is so?

____

____

____

Mixed Applications

4. If Tuesday is the third day of the week, what day is the fourth day?

____

5. What is the value of 5, in the number 157?

____

6. Cindy buys a jacket for $56.25. She gives the store clerk $60.00. What is her change?

____

7. Rupal is at the mall. Her mother tells Rupal to meet her in 10 minutes. If it is 4:30 now, what time should Rupal meet her mother?

____

Name ______________________

Fair or Unfair Games

Vocabulary

Fill in the blank.

1. A game is ________ if everyone has an equal chance at winning.

Choose the box of balls or bag of nuts that is fair. Write *A* or *B*.

2.  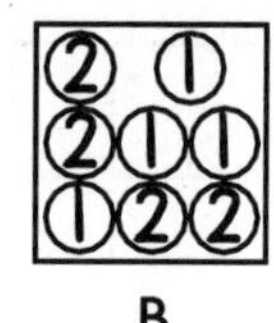 ______

A B

3.   ______

A B

4. 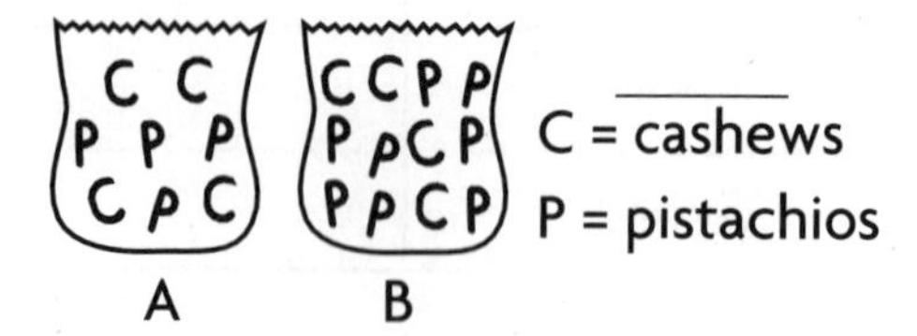

A B ______

5. 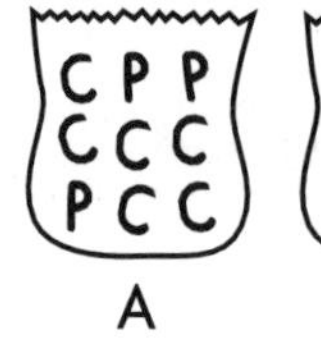

P C C P
P C C
P C P

A B ______

Mixed Applications

6. Choose a fair box of balls from above. Explain how you can make it unfair.

____

7. Sasha has 250 pens. She buys 100 more. How many pens does Sasha have in all?

____

8. The movie starts at 1:15. It is 3 hours long. What time does the movie end?

____

9. I am a number between 100–110. If you subtract 5 tens from me, you get 56. What number am I?

____

10. Write the number in standard form.

a. eight hundred twenty-six

____

b. two thousand four ________

11. Melinda, Elizabeth, and Ann are each writing her name 15 times. Who do you think will finish first? Why?

____

____

Name ______________________________

Sorting and Comparing Solids

Vocabulary

Fill in the blank to complete each sentence.

1. A ______________ is a flat surface of a solid figure.

2. An ______________ is a straight line where 2 faces meet.

3. A ______________ is where 2 or more edges meet.

Complete the table.

	Figure	Faces	Edges	Corners
4.	Cube			
5.	Rectangular prism			
6.	Square pyramid			
7.	Cylinder			
8.	Cone			
9.	Sphere			

Name the solid figure that each looks like.

10. 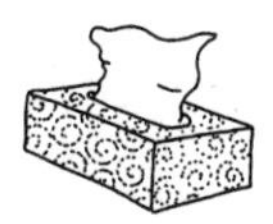

____

11. 

____

12. 

____

13. 

____

14. 

____

15. 

____

Mixed Applications

16. What solid figure has 6 faces that are all alike?

____

17. How many faces do 3 square pyramids have in all?

____

Name ______________________

Tracing and Naming Faces

Write the name of the solid figure that can be made from each of the patterns. You may trace, cut, and fold the figures to check your answers.

1.

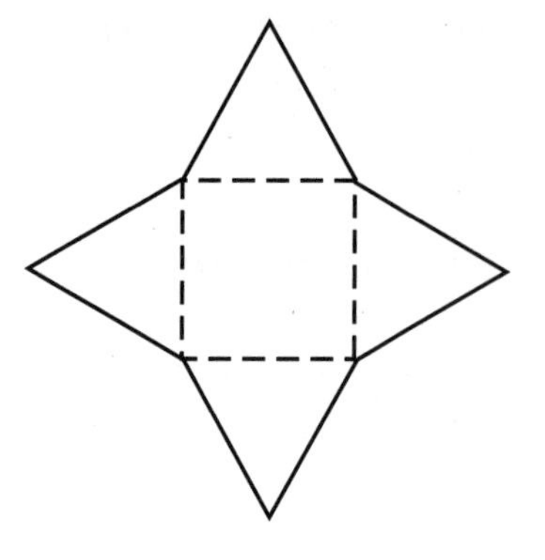

____

2.

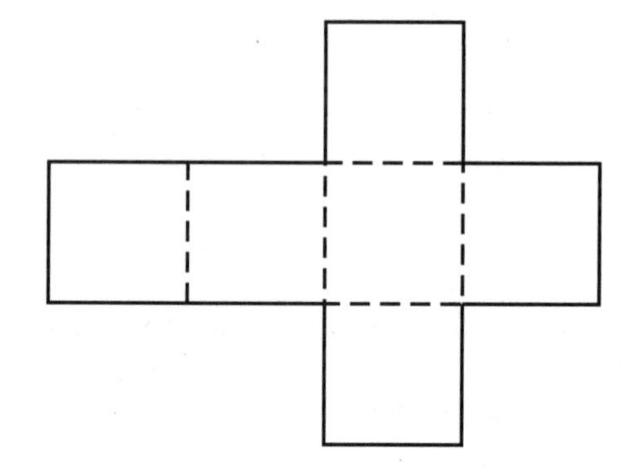

____

3.

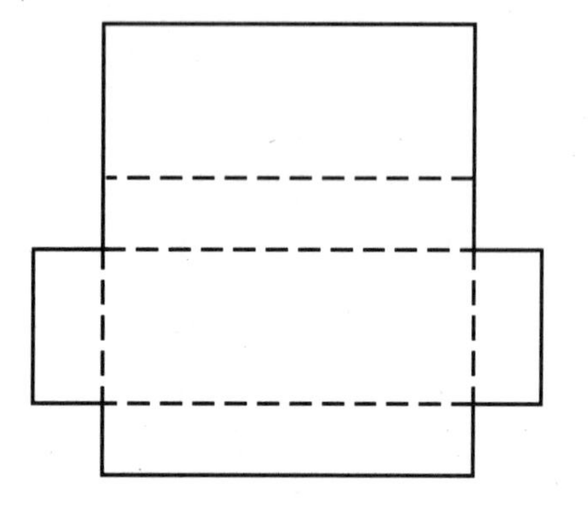

____

4.

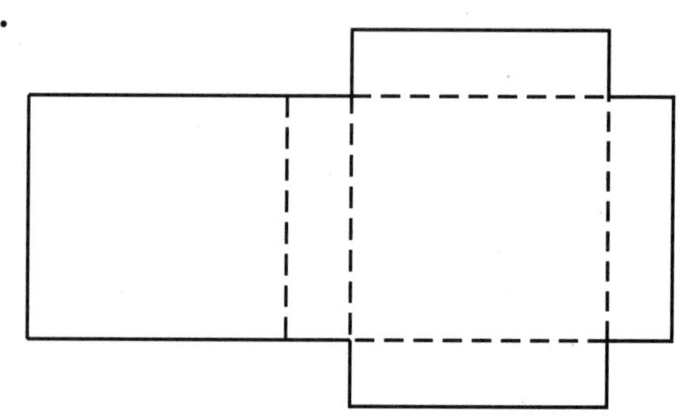

____

Mixed Applications

5. Tim is making a model of a cube. He is using toothpicks for the edges and balls of clay to mark the corners. How many toothpicks does Tim need? how many balls of clay?

____

____

6. Eliza counted the books on her shelf. She said, "I have about 80 books, rounded to the nearest ten." What is the least number of books that Eliza could have? the greatest number?

____

Name ____________________

Matching Faces to Solids

Tell which solid figure has each set of faces.
Circle *a, b,* or *c.*

1. 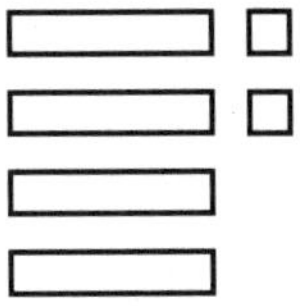

a. rectangular prism
b. cube
c. square pyramid

2. 

a. rectangular prism
b. cube
c. square pyramid

3. 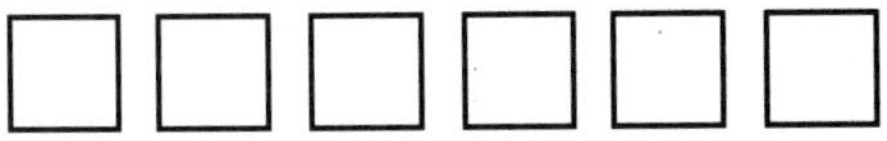

a. rectangular prism
b. cube
c. square pyramid

Mixed Applications

4. Chris stacked 3 cubes. He matched the touching faces of the cubes exactly. What solid figure did he make?

____

5. A parking lot has 6 rows of cars. There are 8 cars in each row. How many cars are in the parking lot?

____

6. Jeremy paid for his lunch with a $5 bill. The cashier gave him $2.39 in change. How much did Jeremy's lunch cost?

____

7. Anne, Carrie, and Steve shared a bag of marbles evenly. There were 27 marbles in all. How many marbles did each person get?

____

8. Linda weighs twice as much as Jim. Mr. Murphy weighs twice as much as Linda. Jim weighs 39 pounds. How much does Mr. Murphy weigh?

____

9. Pete began reading at 10:30. He read 5 pages. Each page took him about 2 minutes to read. At about what time did Pete finish reading?

____

Name ____________________

Plane Figures

Vocabulary

Fill in the blank to complete the sentence.

1. A ____________________ is a flat surface.

2. A ____________________ is a closed figure in a plane.

Tell whether each figure is formed by only straight lines, only curved lines, or both straight and curved lines.

3. 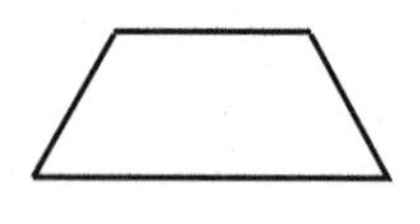
____

4. 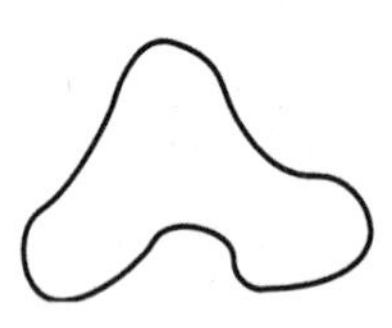
____

5. 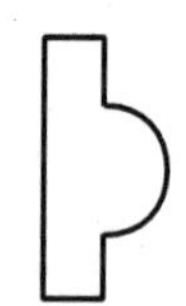
____

6. 
____

7. 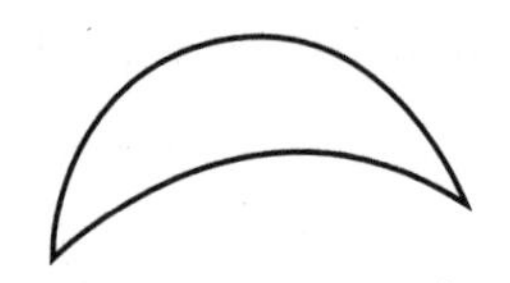
____

8. 
____

9. 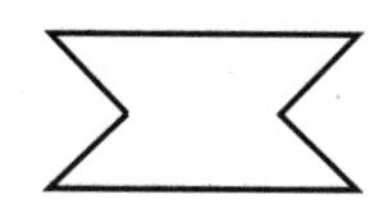
____

10. 
____

Mixed Applications

11. Karen spent 21 days at the beach. How many weeks did she spend at the beach?

____

12. I am a two-digit number. The digit in my ones place is 4 times greater than the digit in my tens place. The sum of my digits is 10. What number am I?

____

13. Which of the following letters are plane figures?
A C D F O R T

____

14. Al has been playing ball for 30 minutes. It is now 1:45. At what time did Al start playing?

____

Name ___________________________________

Patterns with Plane Figures

hexagon parallelogram trapezoid triangle square

Draw the next two shapes in each pattern.

1. ⬡ □ △ □ ⬡ □ △ □ ____ ____

2. ⏢ △ △ □ ⏢ △ △ □ ____ ____

3. ▱ ⬡ □ □ ▱ ⬡ □ □ ____ ____

Tell what shapes are missing in each pattern.

4. ▱ ? □ ▱ △ □ ? △ □

____

5. ⏢ △ ? □ ⏢ ? ⏢ □

____

6. □ □ ? △ □ ? ⏢ △

____

Mixed Applications

7. Lisa is stringing beads. She has started a pattern of sphere, cube, cube, sphere, cube, cube. If Lisa continues this pattern, what kind of bead will she use next?

____

8. Alan shaped a hexagon using 6 toothpicks. How many toothpicks would he need to make 8 hexagons?

____

Name ______________________

LESSON 18.5

Problem-Solving Strategy

Find a Pattern

Find a pattern to solve.

1. Sarah is gluing shapes around a frame. Draw the next three shapes in her pattern.

○ ◇ ○ △ ○ ◇ ___ ___ ___

2. Jeff is decorating the border of a crown. Draw the next three shapes in his pattern.

○ • ● ∘ ○ • ___ ___ ___

3. There is a pattern in the numbers below. What will the next two numbers be?

3, 14, 25, 36, ___, ___

4. Sketch the next two dot triangles to continue the pattern below.

1 3 6 ___ ___

Mixed Applications

Solve.

CHOOSE A STRATEGY

- Guess and Check
- Act It Out
- Make a Model
- Make a List
- Draw a Picture

5. Theo has 30 mice. They are either gray or brown. There are 4 more gray mice than brown mice. How many gray mice are there? brown mice?

____

____

6. Laura is younger than Hans and older than Peggy. Peggy is older than Felix. List the children in order from youngest to oldest.

____

____

7. It is 11:30. Brian just finished playing baseball for an hour. Before that, he raked leaves for 15 minutes. At what time did Brian start raking leaves?

____

8. What are the possible outcomes when you toss two coins?

____

Name ______________________________

LESSON 19.1

Line Segments and Angles

Vocabulary

Write the correct letter from column 2.

1. line _____

2. angle _____

3. line segment _____

a. is straight and is the part of the line between two points

b. is straight and continues in both directions

c. is formed where two line segments cross or meet

Write the number of line segments in each figure.

4. _____

5. _____

6. _____

Write the number of angles in each figure.

7. _____

8. _____

9. _____

Write if each angle is a *right angle, less than* a right angle, or *greater than* a right angle. Use a corner of your paper to help you.

10.

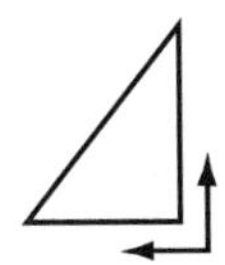

11.

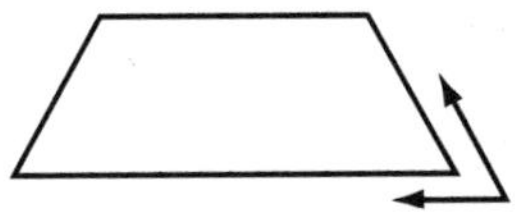

12.

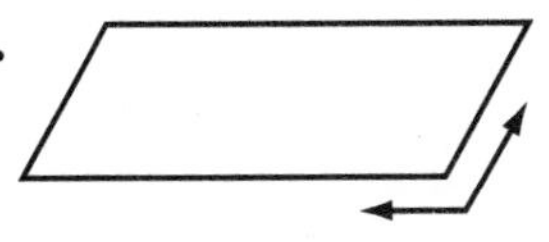

_______________ _______________ _______________

Mixed Applications

13. Marty is the 8th person in line. If Georgina is behind him, what place in line is she?

____

14. Jo shares her 15 pens with her friends. She has 2 friends. How many pens do they each get?

____

Name ______________________________

LESSON 19.2

Locating Points on a Grid

Vocabulary

Fill in the blanks with the correct word.

ordered pair grid

1. A ____________ is a map divided into equal squares.

2. An ____________ of numbers names a point on a grid.

Use the grid. Make a map of the neighborhood. Use the ordered pairs below to place the street names on your map. Record the first letter of the street name on the correct point.

3. (4,5) Rice St.

4. (1,7) Elm St.

5. (6,2) Water St.

6. (2,2) Concord St.

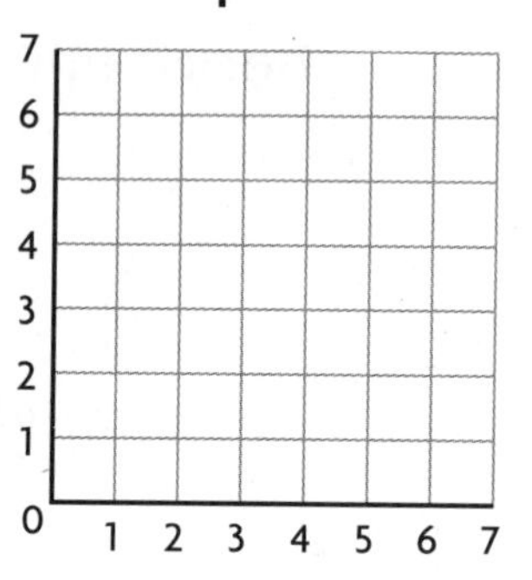

For Exercises 7–12, use the grid. Write each ordered pair for each fruit.

7. apple ______

8. orange ______

9. banana ______

10. grape ______

11. kiwi ______

12. peach ______

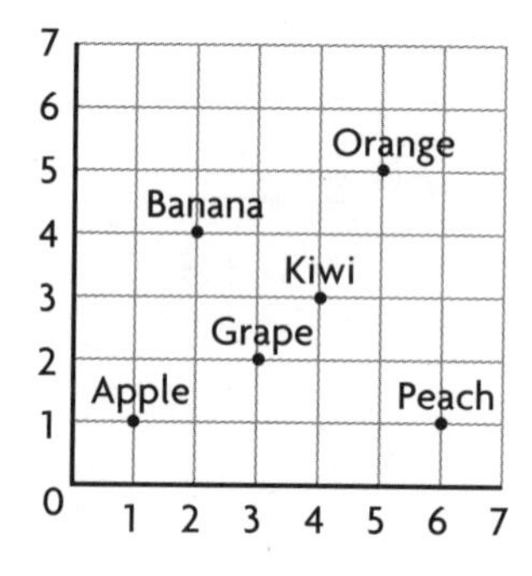

Mixed Applications

13. John's room is a square. How many angles does his room have?

____

14. Rashad picks a fruit at (2,1) on a grid. Ester picks a fruit at (2,3) on a grid. What is another name for the path between these two points?

____

Name ______________________________

Congruent Figures

Vocabulary

Fill in the blank.

1. ______________ figures have the same *size* and *shape*.

---

2. Compare figures A and B. Are the figures congruent? Explain.

A B

____

____

____

3. Compare figures C and D. Are the figures congruent? Explain.

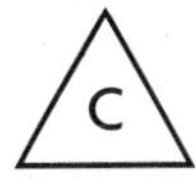

____

____

____

Tell whether the two figures are congruent. Write *yes* or *no.*

4. ________

5. ________

6. 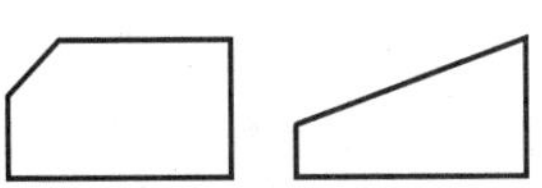

____

Mixed Applications

7. I am a figure with 4 sides and 4 right angles. All my sides are equal. What figure am I?

____

8. Conor buys a pencil for $0.59. He hands the clerk $1.00. What is his change?

____

9. A spinner has the numbers 2, 4, 6, 8, and 10 on it. Is it certain or impossible that you will spin an odd number?

____

10. The minute hand on the clock shows 27 minutes after the hour. How many minutes will it take to get to the next hour?

____

Name ______________________________

Using Congruent Figures

Trace and cut out figures A through D. Use the cutout figures to make a design in the box below. Count the number of congruent figures used in your design. Record the numbers in a table.

Shape	Number of Congruent Figures
A	
B	
C	
D	

Mixed Applications

1. There are 16 pencils. There are 8 students. How many pencils does each student get?

   ____

2. Mark goes to school at 9:00 A.M. He gets out at 3:15 P.M. How long is he at school?

   ____

3. Amy is behind Robin in line at the movies. Marcy is in front of Amy, and Heidi is behind Amy. If Robin is first in line, who is second, third, and fourth in line?

   ____

   ____

4. Julia and Sam walk to and from school. It takes them 20 minutes. How many minutes do they walk daily?

   ____

Name ______________________________

Congruent Solid Figures

For Exercises 1–3, use cubes to build each figure. Write how many cubes you need for each figure.

1. 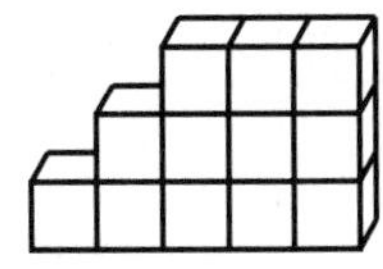

2. 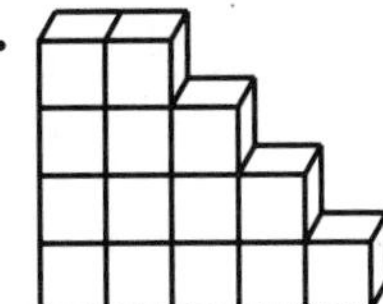

3. 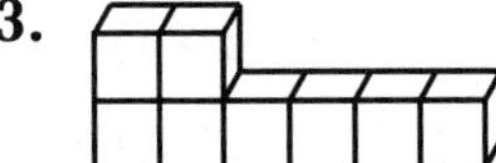

4. Are the solid figures in Exercises 1 and 2 congruent? Explain.

Write the letter of the solid figure that is congruent with the first solid figure.

5. 

A. 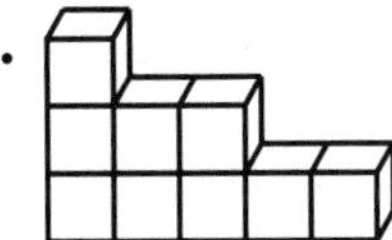

B. 

Mixed Applications

6. The numbers for the first four offices are 124, 126, 128, and 130. What are the numbers for the fifth and sixth offices?

7. There are 3 bags. In each bag there are 5 candles. How many candles are there in all?

8. Joan and Ed have 14 sheets of paper. How many sheets do they each get if they divide the paper equally?

9. Maya has two $1 bills, 3 quarters, 4 dimes, 6 nickels, and 12 pennies. How much money does she have in all? Does she have enough to buy a book for $4.00?

Name ____________________

Problem-Solving Strategy

Make a List

Make a list to solve.

1. Suzanne made these solid figures with connecting blocks. Are they congruent? Explain your answer.

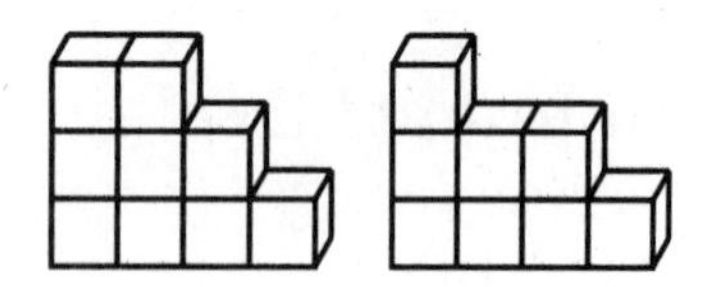

____

____

____

2. Scott built a solid figure with 5 blocks in the first layer, 3 blocks in the second layer, and 2 blocks in the third layer. Is this figure the one Scott made?

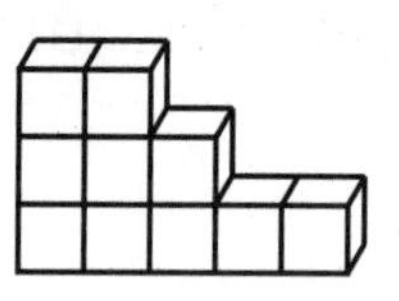

____

____

____

Mixed Applications

Solve.

CHOOSE A STRATEGY

- **Guess and Check**
- **Draw a Picture**
- **Find a Pattern**
- **Write a Number Sentence**

3. Five people want to play a marble game. Kim has 25 marbles to share equally. How many marbles does each person get?

____

4. Bonnie went to Maria's house. She arrived at 3:30 and stayed until 6:00. How long was Bonnie at Maria's house?

____

5. Corey practiced skating every day for 21 days. For how many weeks did she practice?

____

6. Charlie swims 12 minutes longer than Tony. They swim a total of 50 minutes. How many minutes does each swim?

____

____

Name ______________________

Sliding, Flipping, and Turning

Fill in the blank with a word from the Word Box.

Word Box
flip
slide
turn

1. You ____________ a figure when you move it in a straight line.

2. You ____________ a figure when you move it over a line.

3. You ____________ a figure when you move it around a point.

Tell what kind of motion was used to move each plane figure. Write *slide*, *flip*, or *turn*.

4. 

____

5. 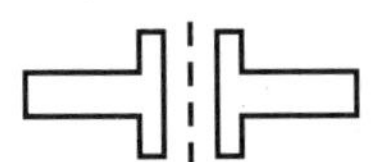

____

6. 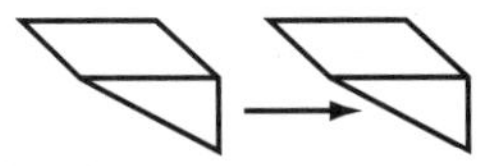

____

7. 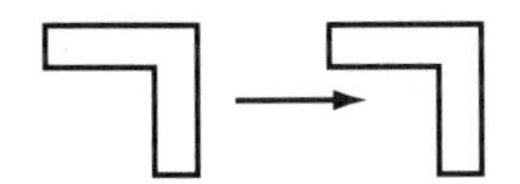

____

8. 

____

9. 

____

Mixed Applications

10. Jeff is sliding a coin from one side of a board to another side. Will the slide change the size of the coin? Explain.

____

____

11. Max is flipping a playing card over a line. Will the flip change the shape of the playing card? Explain.

____

____

12. Tim has 24 baseball cards. He puts them into stacks of 4. How many stacks of 4 cards does he have?

____

13. Rick was 20 minutes early for a basketball game. The game started at 7:30. At what time did Rick arrive?

____

Name ______________________________

Symmetry

Fill in both blanks with the same group of words.

1. A ______________________ is an imaginary line that divides a figure in half. If you fold a figure along a ______________________, the two sides match.

Is the broken line a line of symmetry? Write *yes* or *no*.

2. 

____

3. 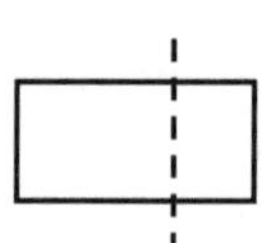

____

4. 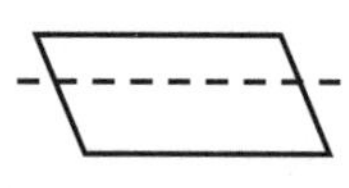

____

5. 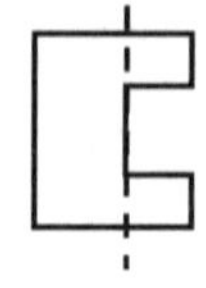

____

6. 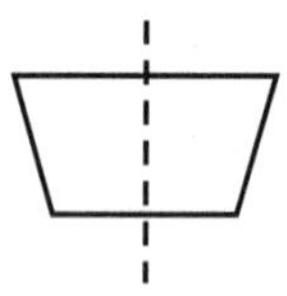

____

7. 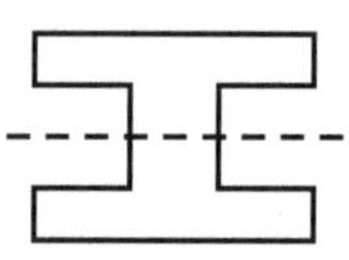

____

Mixed Applications

8. Draw a square. Show a line of symmetry with a dotted line.

9. If you were to flip the letter Y over a line running under it, what would it look like?

____

10. Ted has 36 baseball cards. He gives 8 cards to his sister. Then he divides the rest between himself and his friend. How many cards does his friend get?

____

11. Rod shared a package of 40 lollipops equally with 7 friends. How many lollipops did Rod and each of his friends get?

____

Name ______________________________

More About Symmetry

Is the broken line a line of symmetry? Write *yes* or *no*.

1. 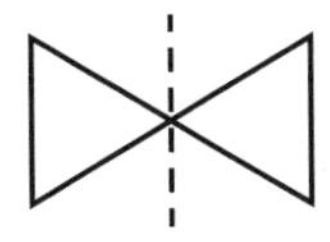
____

2. 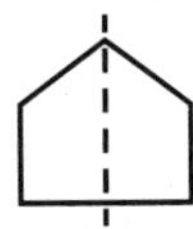
____

3. 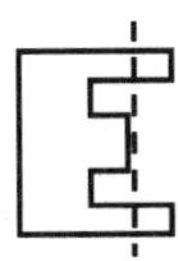
____

Draw the lines of symmetry. How many lines of symmetry does each figure have?

4. 
______

5. 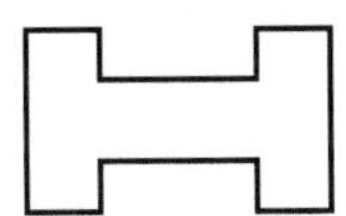
______

6. 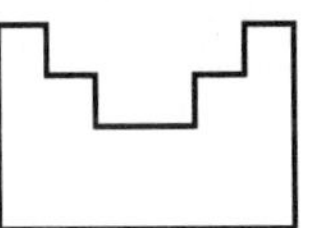
______

Mixed Applications

7. Write a letter that has 1 line of symmetry. Then draw the line.

8. Write a letter that has 2 lines of symmetry. Then draw the lines.

9. Paul made 81 points on his third math test. This is 6 points lower than his score on his first math test and 11 points higher than his score on his second math test. What is the difference between his scores on his first and second tests?

____

10. Erica collected 38 cans this week. Last week, she collected 7 fewer cans. How many cans did she collect all together?

____

Name ______________________________

LESSON 20.4

Symmetric Patterns

1. How is symmetry like a mirror image?

____

____

____

Complete each drawing to make a symmetric figure.

2. 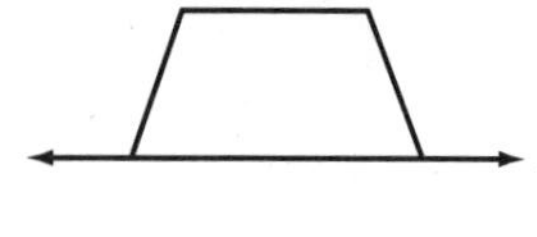

3. 

4. 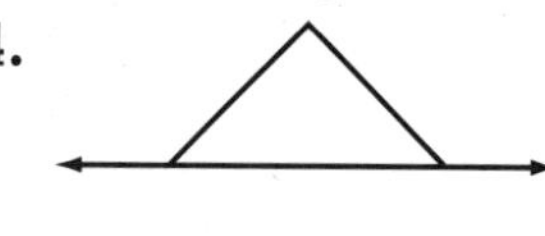

5. 

6. 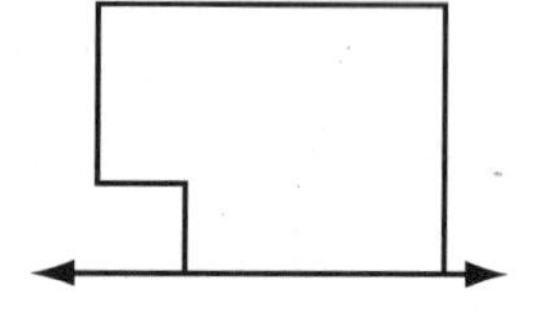

7. 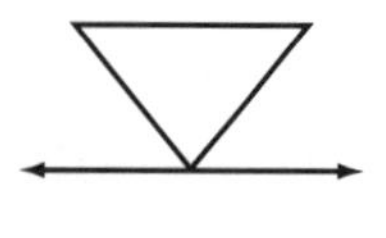

8. 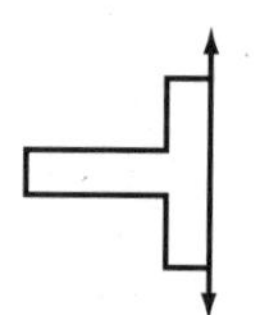

9. 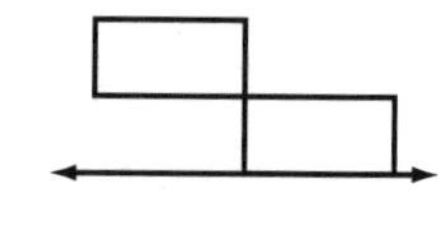

Mixed Applications

10. What farm animal has a name that looks the same even if you flip it over a line under it? (HINT: two letters)

____

11. Vinny and Tony have 2 daily chores each. In all, how many chores do they do in 3 weeks?

____

12. Ursula has supper 3 hours and 15 minutes after she gets home from school at 3:30. What time is supper served?

____

13. Juanita wants to buy a movie video that costs $14.95. She baby-sits for 3 hours, earning $3.50 an hour. How much more money does she need?

____

Name ______________________________

Problem-Solving Strategy

Draw a Picture

Draw a picture to solve.

1. Anita made this design. Byron matched it to make the other half. How did it look when Byron was finished?

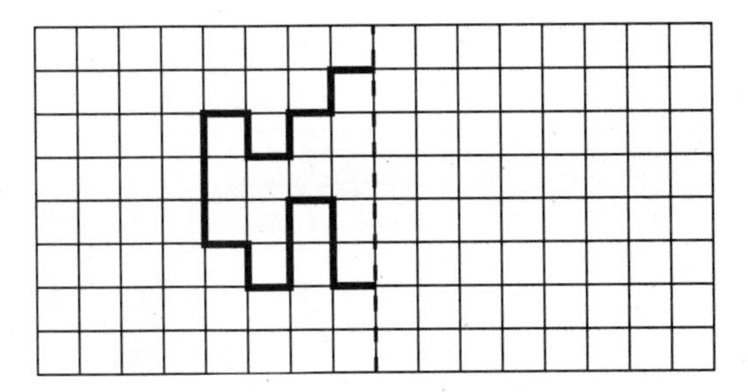

2. The front of Meg's house could be divided by a line of symmetry. There are 8 windows and 4 columns. How many windows and columns are on each side of the line of symmetry? Where is the front door?

____

____

3. Mrs. Simmons lines her cast members in a symmetrical pattern on stage. The 2 stars are in the middle. There is 1 dancer on each side of the stars and 2 singers at each end of the line-up. How does the line-up look from one end to the other?

____

____

____

4. To go to the library, Becky turns left when she leaves home and walks 3 blocks. Then she turns right and walks 2 blocks. Then she turns right for 1 block and turns left for another 2 blocks. Draw her route.

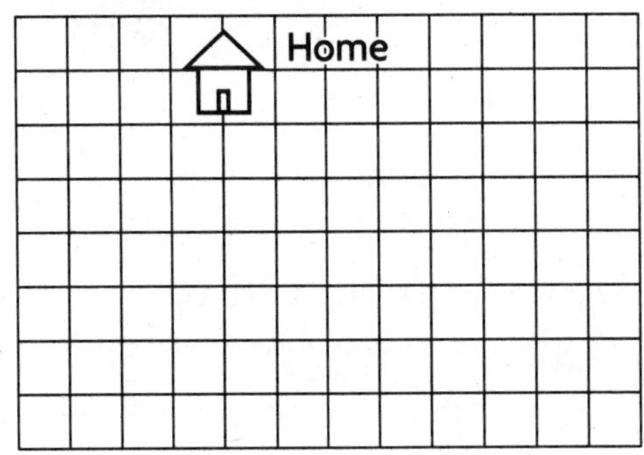

Mixed Applications

5. Write a number between 795 and 810 that can be divided by a line of symmetry through the middle digit.

____

6. Nate spent $3.29 of his allowance and had $2.21 left. How much is his allowance?

____

Name ______________________________

LESSON 21.1

Modeling Parts of a Whole

Tell how many parts make up the whole. Then tell how many parts are shaded.

1. 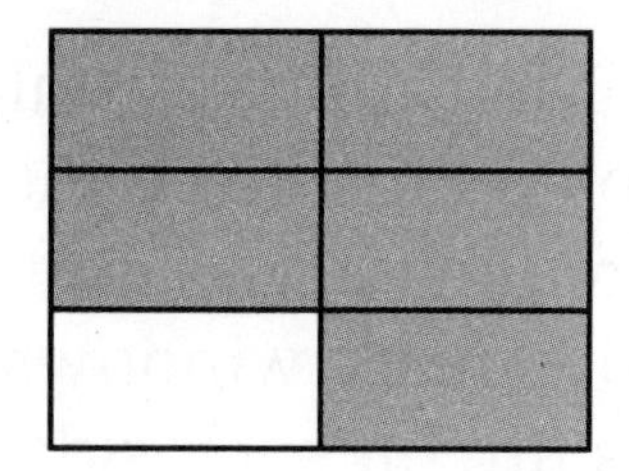

____

2. 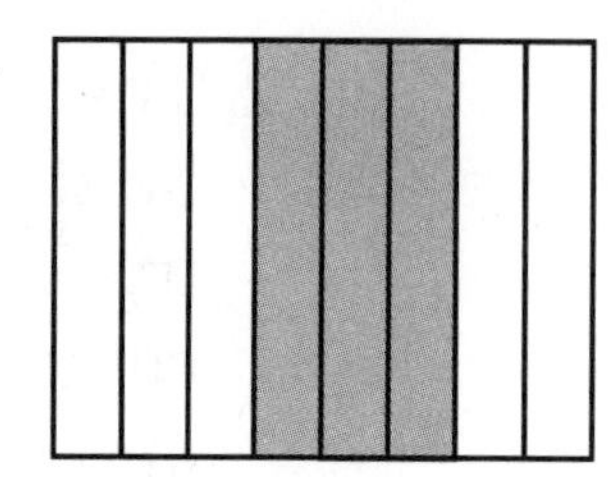

____

3. 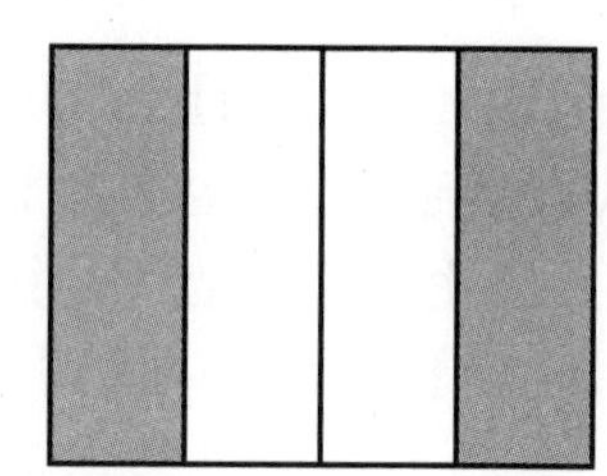

____

4. 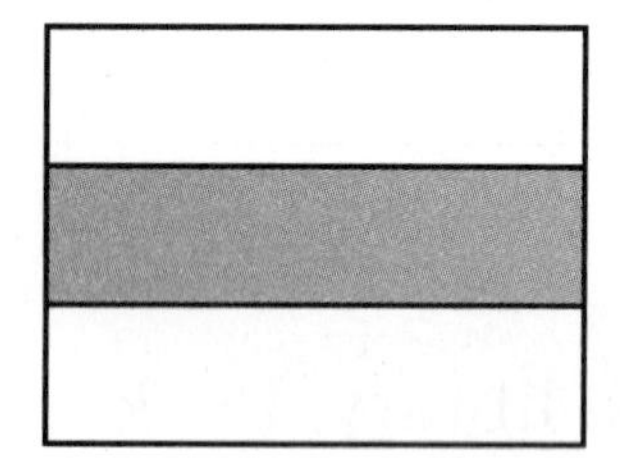

____

5. 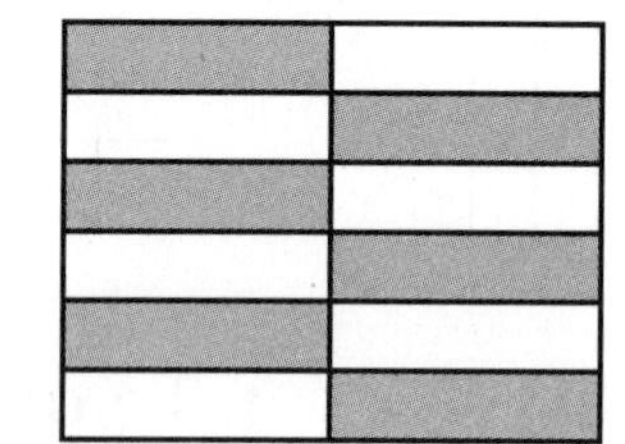

____

6. 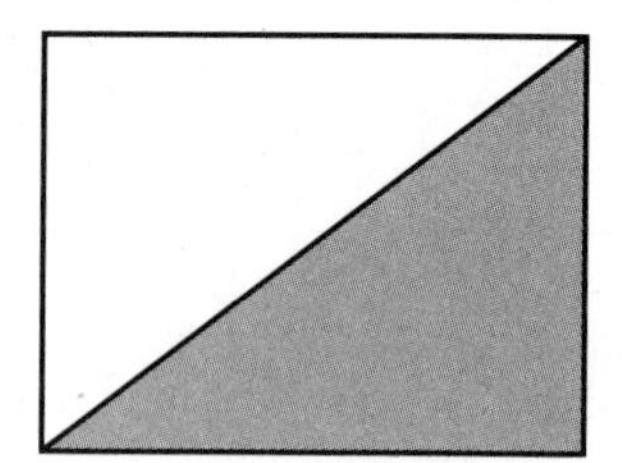

____

Mixed Applications

7. Don cut a pizza into 8 slices. How many slices make up the whole pizza? If he eats 2 slices, how many parts are eaten?

____

8. Jamal's golf class starts in 3 weeks. Today is July 5. On what date does the class start?

____

9. Amy cut a cake into 10 slices. How many slices make up the whole cake? If she gives 4 slices to her friends, how many parts are missing from the cake?

____

10. Tom and Al started jogging at 10:15. They jogged together for an hour. Then Tom continued alone for 12 minutes more. At what time did Tom stop jogging?

____

Name ______________________________

Other Ways to Model Fractions

Match each word in Column 1 with its definition in Column 2.

Column 1	Column 2
_____ 1. denominator	a. tells how many parts are being used
_____ 2. fraction	b. tells how many equal parts are in a whole
_____ 3. numerator	c. a number that names part of a whole

Tell the part that is shaded. Write your answer using numbers and words.

4. 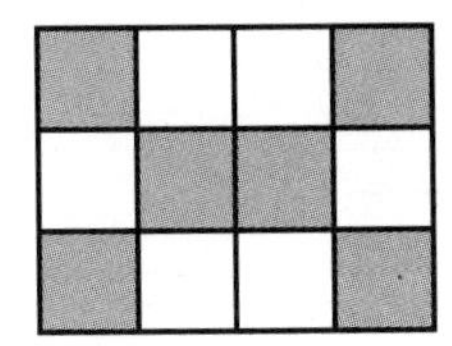

____

5. 

____

6. 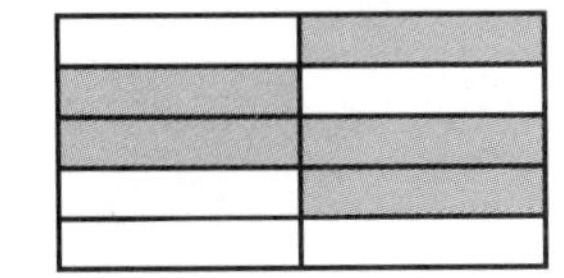

____

Write the fraction using numbers.

7. three fifths _____

8. six out of eleven _____

9. two divided by three _____

10. one out of six _____

11. nine divided by ten _____

12. seven twelfths _____

Mixed Applications

13. Will spent $1.25 on pens, $4.90 on a book, and $2.35 on paper. How much did he spend in all?

____

14. May bought a dozen eggs. She used 5 eggs while baking pies. What part of the dozen did she use?

____

Name ______________________

LESSON 21.3

Counting Parts to Make a Whole

Write a fraction to describe each shaded part.

1. 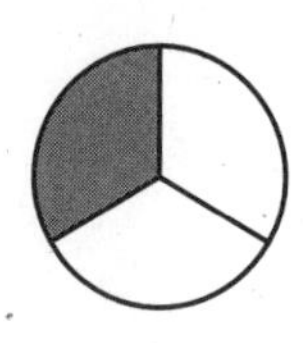 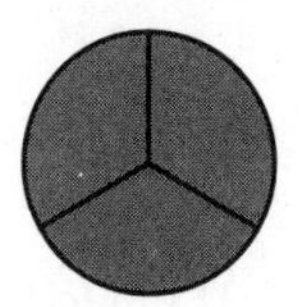

____

2. 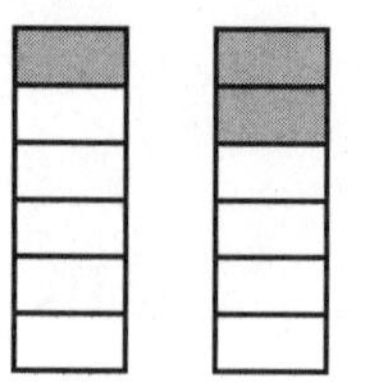 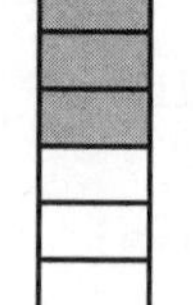 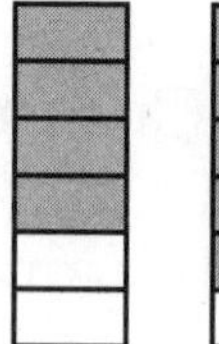 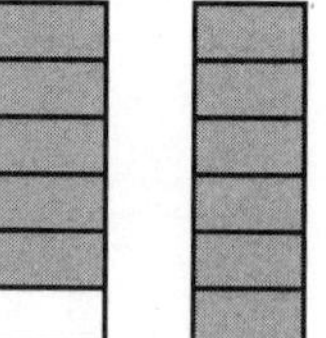

____

3. 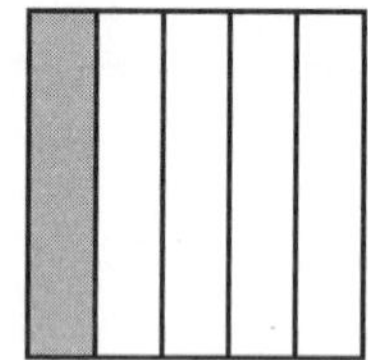 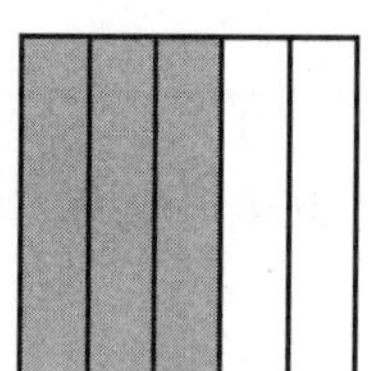 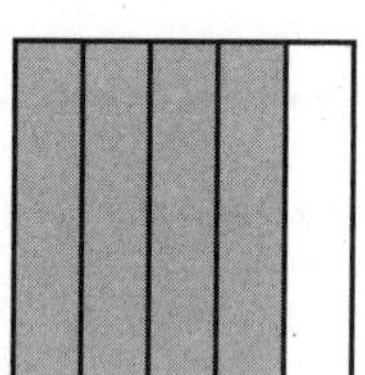 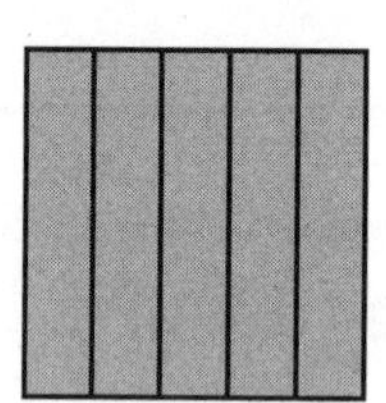

____

Write a fraction that names the shaded part.

4. 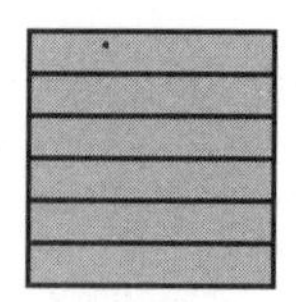

______

5. 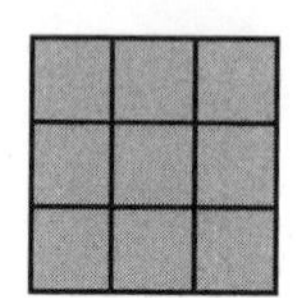

______

6. 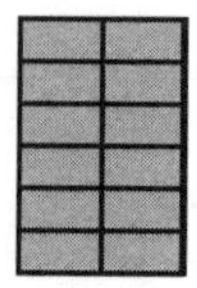

______

Mixed Applications

7. What fractional part of the figure is shaded? not shaded?

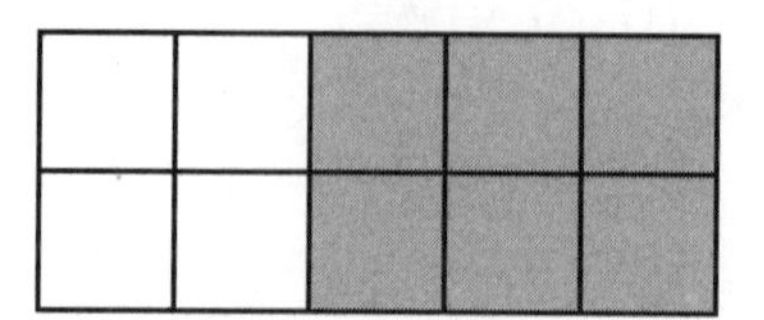

____

8. Marissa started her homework at 6:30. She finished $2\frac{1}{2}$ hours later. What time was it then?

____

9. Stacey earned $10 two weeks ago, $5 last week, and $10 this week. If this pattern repeats for 3 more weeks, how much will she have in all?

____

Name ______________________________

LESSON 21.4

Comparing Fractions

Compare. Write <, >, or = in each ◯.

1.

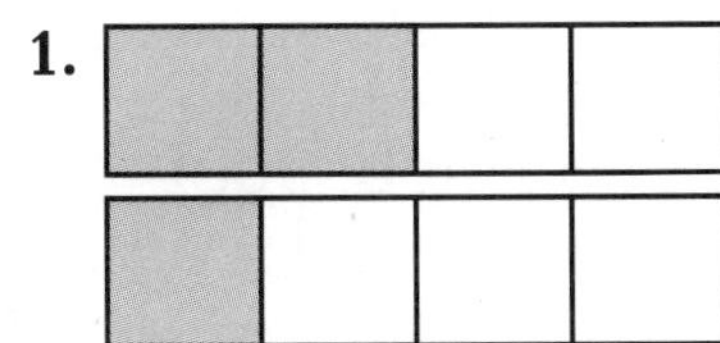

$\frac{2}{4}$ ◯ $\frac{1}{4}$

2.

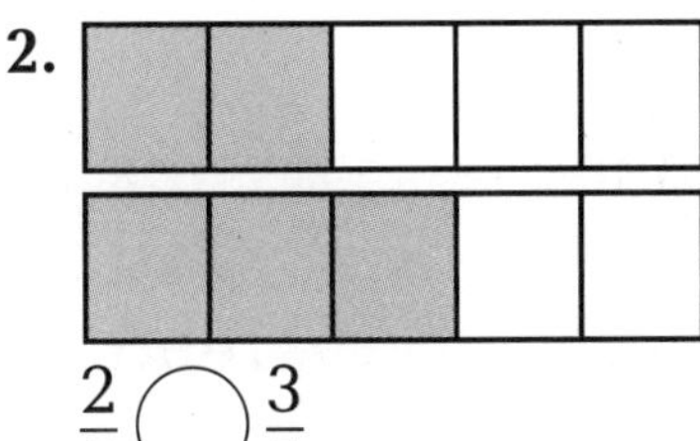

$\frac{2}{5}$ ◯ $\frac{3}{5}$

3.

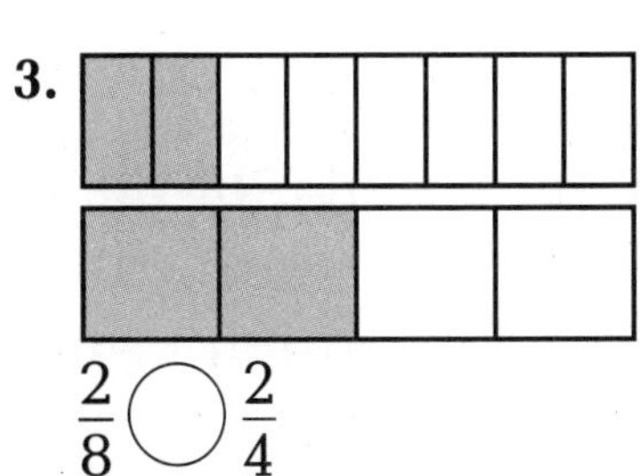

$\frac{2}{8}$ ◯ $\frac{2}{4}$

4.

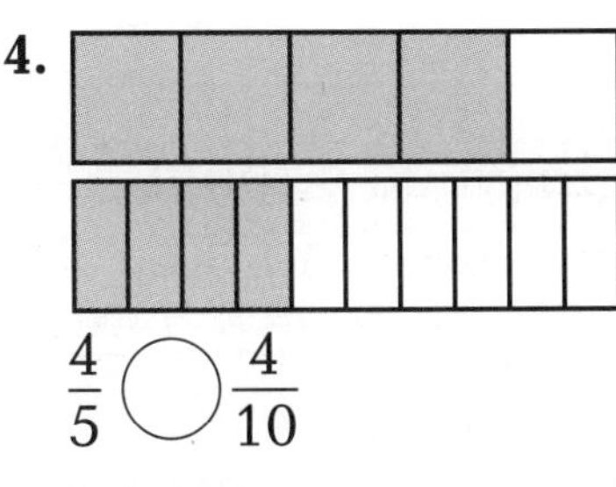

$\frac{4}{5}$ ◯ $\frac{4}{10}$

5.

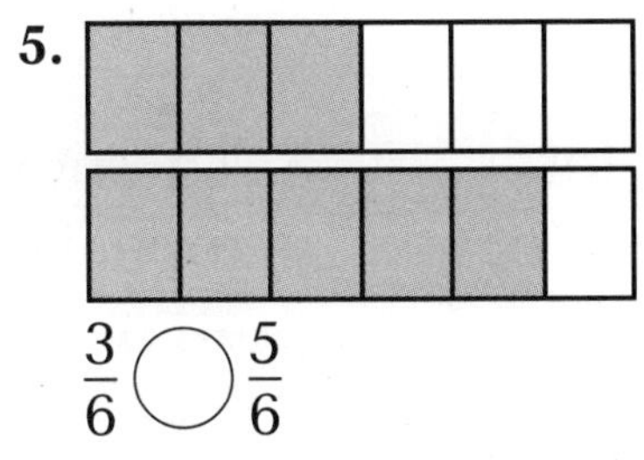

$\frac{3}{6}$ ◯ $\frac{5}{6}$

6.

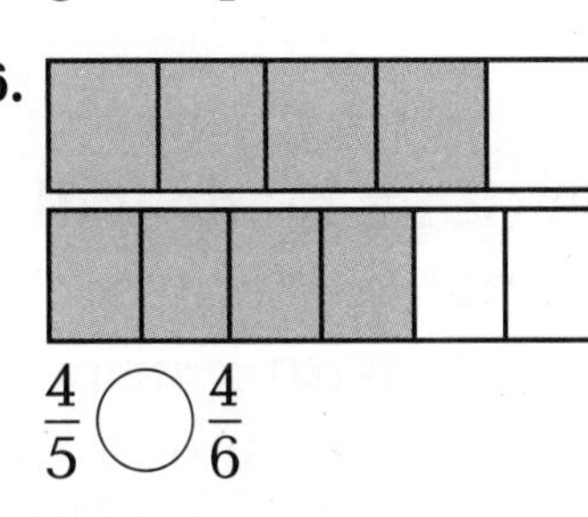

$\frac{4}{5}$ ◯ $\frac{4}{6}$

7.

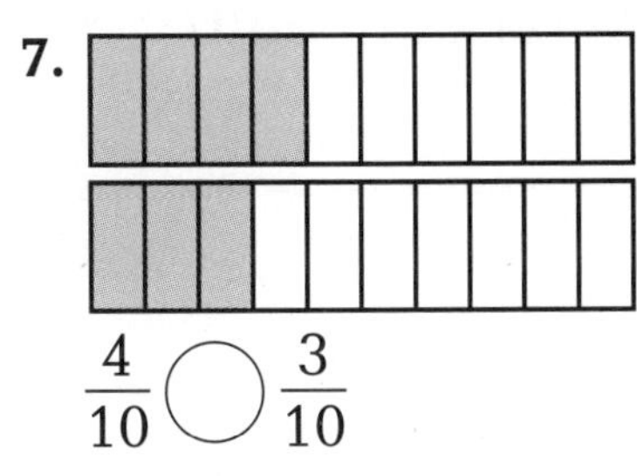

$\frac{4}{10}$ ◯ $\frac{3}{10}$

8.

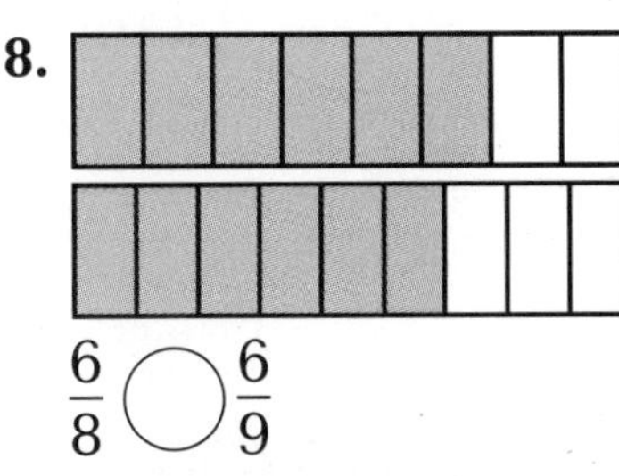

$\frac{6}{8}$ ◯ $\frac{6}{9}$

9.

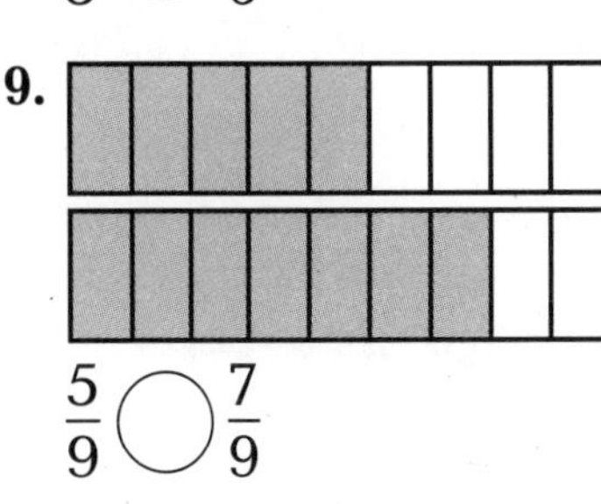

$\frac{5}{9}$ ◯ $\frac{7}{9}$

10.

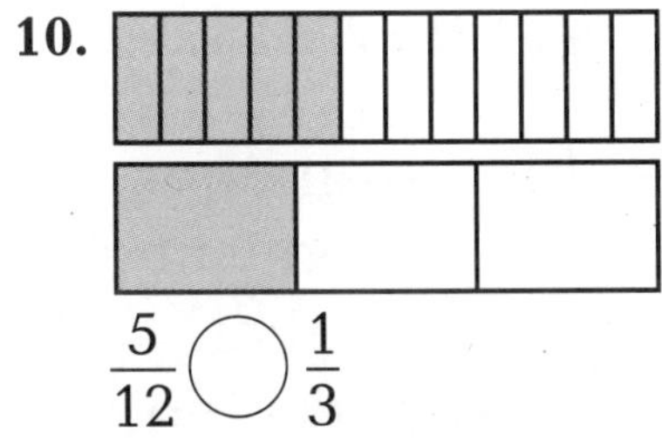

$\frac{5}{12}$ ◯ $\frac{1}{3}$

11.

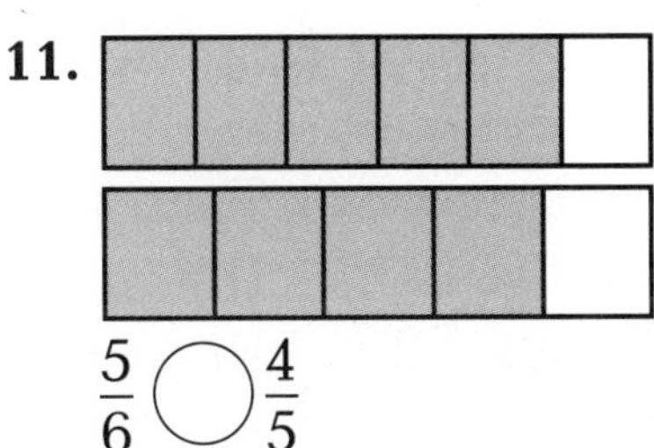

$\frac{5}{6}$ ◯ $\frac{4}{5}$

12.

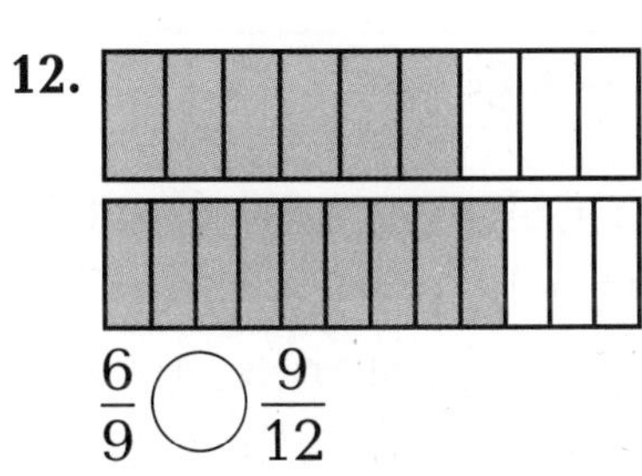

$\frac{6}{9}$ ◯ $\frac{9}{12}$

Mixed Applications

13. Jill began reading at 10:30. She read for 50 minutes. Then she rode her bicycle for 35 minutes. What time is it now?

____

14. Find the next three numbers in this pattern: 5, 15, 20, 30, 35.

____

15. An egg carton holds 12 eggs. How many cartons are needed to hold 48 eggs?

____

16. Todd read $\frac{4}{7}$ of a book. Sam read $\frac{1}{2}$ of the same book. Who read more?

____

Name ______________________________

LESSON 21.4

Problem-Solving Strategy

Draw a Picture

Draw a picture to solve.

1. Sean spent $\frac{3}{7}$ of his allowance on a book and $\frac{2}{5}$ on a baseball. On which item did he spend more?

____

2. Alex read $\frac{3}{8}$ of a book. Joel read $\frac{2}{5}$ of the same book. Who read more?

____

3. Mr. Ruiz made a divider for his patio. He used 9 stacks of bricks, with 7 bricks in each stack. How many bricks did he use?

____

4. The border in Shea's room repeats 2 triangles and a circle after each square. If one wall has 9 repeats, how many triangles are on that wall?

____

Mixed Applications

Solve.

CHOOSE A STRATEGY

- Draw a Picture • Act It Out • Make a Model • Find a Pattern • Write a Number Sentence

5. Tia, Juan, and Carla are standing in a line. Tia is behind Juan. Carla is in front of Juan. In what order are they standing?

____

6. Pam used $\frac{1}{4}$ of a dozen eggs for breakfast. She used $\frac{1}{6}$ of a dozen eggs for lunch. For which meal did she use the larger part of the dozen?

____

7. Nancy is getting in shape. Every morning she exercises for 1 hour, and then she rests for 30 minutes; then she repeats that pattern. If she starts exercises at 8:30, what will she be doing at 10:45?

____

8. There are 67 marbles in a jar. Ed takes out 22 marbles on Monday. On Tuesday, Ed puts 35 marbles into the jar. How many marbles are in the jar now?

____

Name ______________________________

Equivalent Fractions

1. Two or more fractions that name the same amount are

______________________________.

Find an equivalent fraction for each shaded part. Use fraction bars to help you.

2. 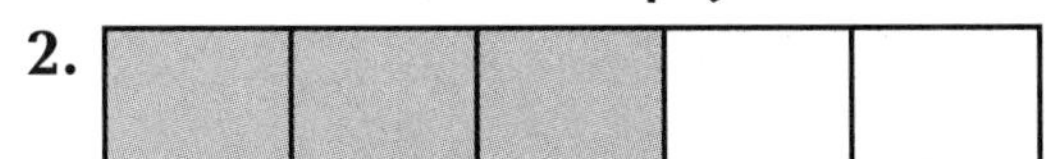

____

3.

____

Find an equivalent fraction. Use fraction bars to help you.

4. $\frac{1}{3} = \frac{\square}{6}$

5. $\frac{3}{5} = \frac{\square}{15}$

6. $\frac{3}{4} = \frac{\square}{16}$

7. $\frac{1}{10} = \frac{\square}{20}$

8. $\frac{12}{12} = \frac{\square}{20}$

9. $\frac{2}{3} = \frac{\square}{12}$

10. $\frac{1}{8} = \frac{\square}{24}$

11. $\frac{5}{7} = \frac{\square}{14}$

Mixed Applications

For Problems 12–13, use the recipe.

12. Zack has only a $\frac{1}{3}$-cup measure. How many $\frac{1}{3}$-cup measures should he use to measure the melon? To measure the strawberries?

____

13. Order the ingredients in amounts from least to greatest.

____

____

FRUIT SURPRISE

$1\frac{1}{3}$ cups melon

$\frac{2}{3}$ cup cherries

$2\frac{2}{3}$ cups strawberries

$1\frac{2}{3}$ cups grapes

Name ________________________________

Part of a Group

Use tiles to show equal parts of the group. Draw a picture of your model.

1. Make 3 equal parts with 1 part green.

2. Make 5 equal parts with 1 part blue.

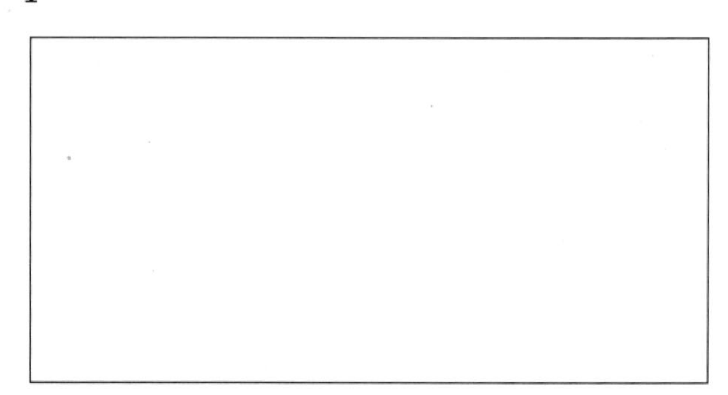

Look at the picture. Find the number of equal parts that are shaded.

3.

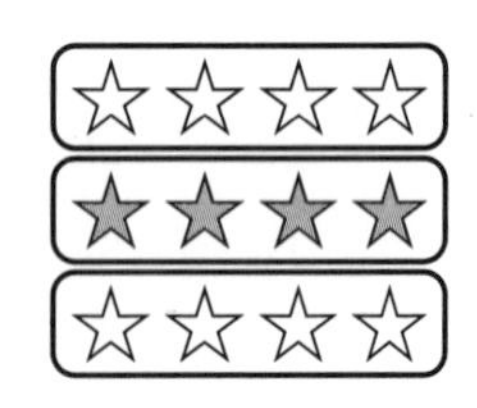

Number of equal parts: ____

Number of parts shaded: ____

4.

Number of equal parts: ____

Number of parts shaded: ____

Mixed Applications

5. Joe and Todd bought a pizza that had 8 slices. They ate all but one slice. What fraction names the part of the pizza that they ate?

____

6. Andrea opened her reading book and saw that the sum of the facing page numbers was 49. What were the two page numbers she opened to?

____

7. How many rectangles are in this figure? Be careful, there are more than 4!

____

Name ______________________________

LESSON 22.2

Fractions of a Group

Write the fraction that names the part of the group shown.

1. 

squares ____

2. 

cats ____

3.

footballs ____

4.

black marbles ____

Draw the picture. Use numbers and words to describe the part that is shaded.

5. Draw 4 rectangles. Shade 1 rectangle.

____

6. Draw 5 circles. Shade 1 circle.

____

7. Draw 9 squares. Make 3 equal groups. Shade 1 group.

____

8. Draw 6 triangles. Make 2 equal groups. Shade 1 group.

____

Mixed Applications

9. Chelsea picked 12 apples. She divided them into 3 equal groups. She made a pie with 1 group. What part of the apples did Chelsea use to make a pie?

____

10. What part of the rectangle is shaded?

____

Name ______________________________

More About Fractions of a Group

Use a pattern to complete the table.

1.	Model	○ ○ ○	● ○ ○	● ● ○	
2.	Number of parts	3		3	3
3.	Number of shaded parts		1	2	3
4.	Fraction of shaded parts	$\frac{0}{3}$	$\frac{1}{3}$		$\frac{3}{3}$

Write a fraction to describe the shaded part.

5. ______

6. ______

7.

______

8.

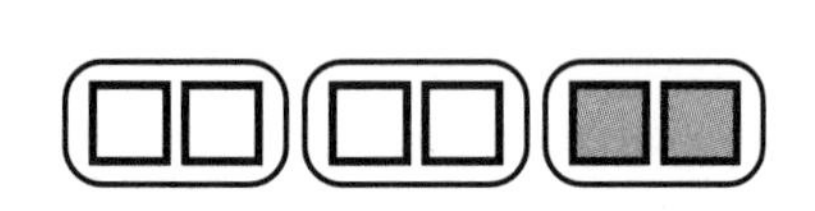

______

9.

______

10.

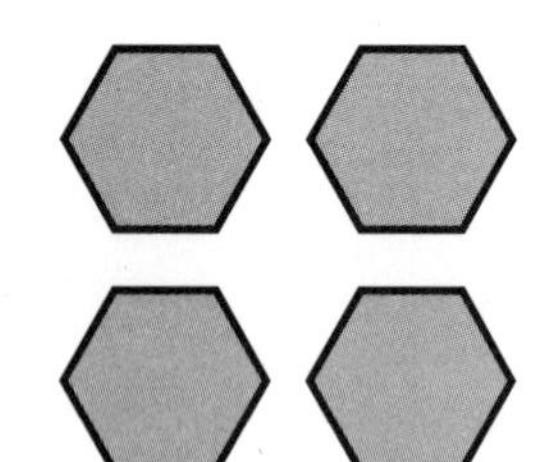

______

Mixed Applications

11. Mrs. Reed has 3 black pens, 1 blue pen, and 3 red pens on her desk. What part of the pens are red?

____

12. If Ted is 8 years old when his father is 35 years old, how old will Ted's father be when Ted is 21 years old?

____

Name ______________________________

Problem-Solving Strategy

Draw a Picture

Draw a picture to solve.

1. There is a litter of 10 puppies at the pet store. Mr. Jones sells 5 puppies. What part of the litter does Mr. Jones sell?

____

2. Eric cuts a cake into 16 pieces. He takes 4 of the pieces on a picnic. What part of the cake does Eric take on the picnic?

____

3. Martin, Judy, and Elizabeth share a package of 12 small puzzles equally. What part of the package does each person get?

____

4. A container holds 5 cups of water when it is full. Tom pours 4 cups of water into the empty container. What part of the container is full? empty?

____

Mixed Applications

Solve.

CHOOSE A STRATEGY

- **Make a Model** • **Guess and Check** • **Act It Out** **Use a Table** • **Work Backward**

5. Alice and Sam are twins. Their brother Mike is 4 years older than they are. If you add the ages of all 3 children, you get 28. How old are the twins?

____

6. Caleb buys a book and a bookmark for $4.14. The bookmark costs $0.89. What is the cost of the book?

____

7. Of the 12 children on the playground, $\frac{1}{4}$ are swinging and the rest are playing tag. Are there more children swinging or playing tag?

____

8. Anne thought of a number. When she added 3 to the number and then doubled the new number, she got 14. What number did she think of first?

____

Name ______________________________

LESSON 22.4

Comparing Parts of a Group

Compare the shaded parts of each group.
Write <, >, or = in the ○.

1. 

$\frac{2}{5}$ ○ $\frac{4}{5}$

2. 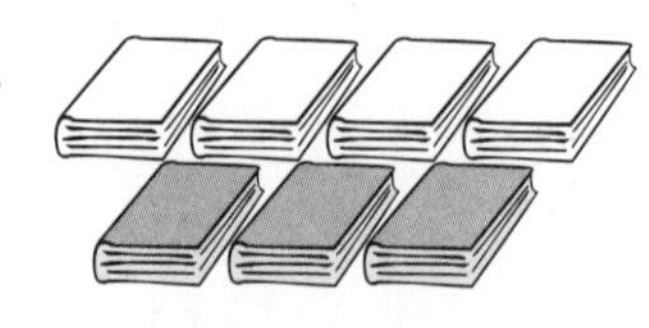

$\frac{3}{7}$ ○ $\frac{2}{7}$

3. 

$\frac{3}{4}$ ○ $\frac{4}{4}$

4. 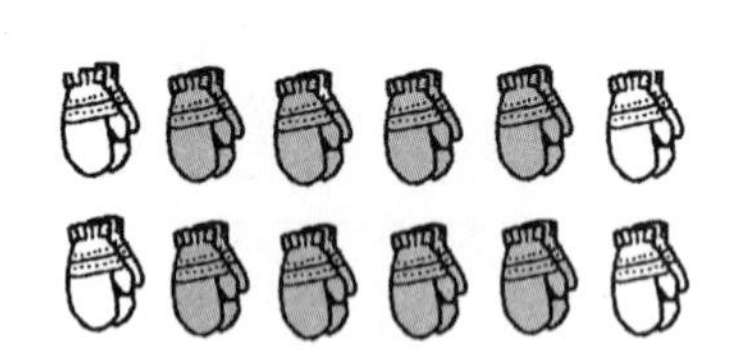

$\frac{4}{6}$ ○ $\frac{4}{6}$

5. 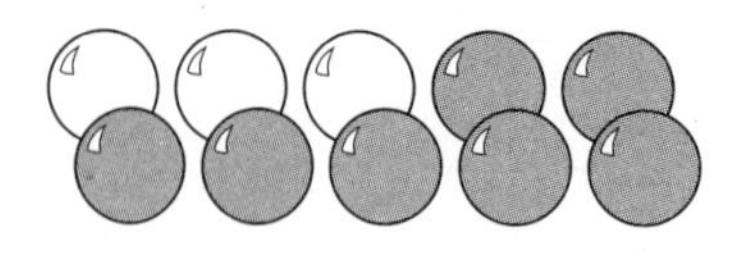

$\frac{3}{10}$ ○ $\frac{7}{10}$

6. 

$\frac{1}{3}$ ○ $\frac{2}{3}$

Mixed Applications

7. Brent put 8 flowers in a vase. Of the 8 flowers, $\frac{3}{8}$ are yellow and $\frac{5}{8}$ are red. Are there more red or yellow flowers?

____

8. Jerry has $0.45 in dimes and nickels. He has the same number of dimes as he has nickels. How many of each coin does he have?

____

9. A group of 45 students formed teams with 9 students on each team. How many teams did the group form?

____

10. Amanda cut her sandwich into 4 equal parts. She ate 2 of the 4 parts. Write two equivalent fractions that name the part of the sandwich that Amanda ate.

____

Name ______________________________________

Tenths

Vocabulary

Fill in the blank to complete each sentence.

1. One of ten equal parts is a ____________.

2. A ____________ is a number that uses place value and a decimal point to show amounts that are less than one, such as tenths.

Write the decimal and fraction for the part that is shaded.

3. 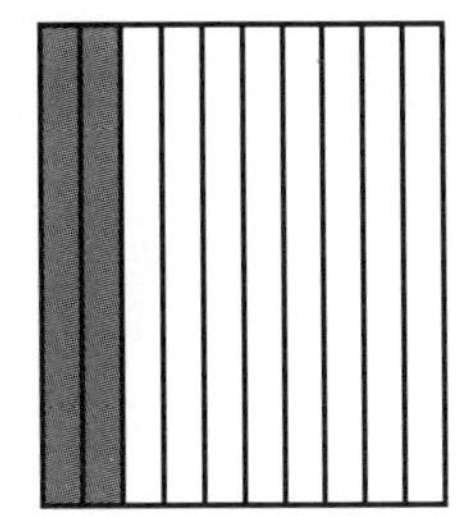
____

4. 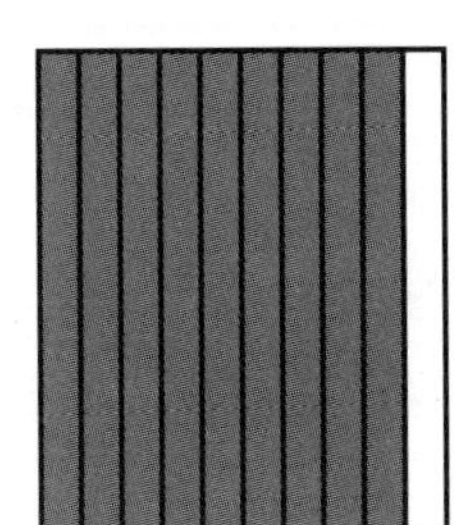
____

5. 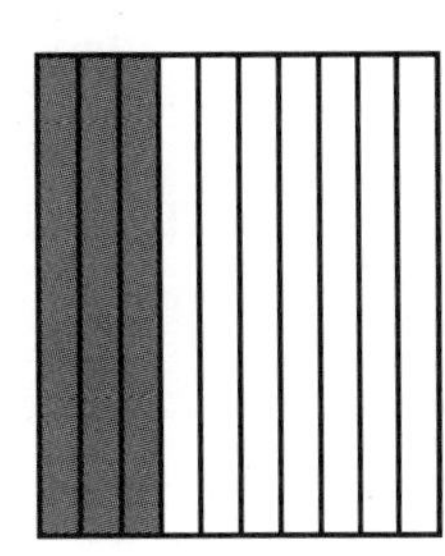
____

6. 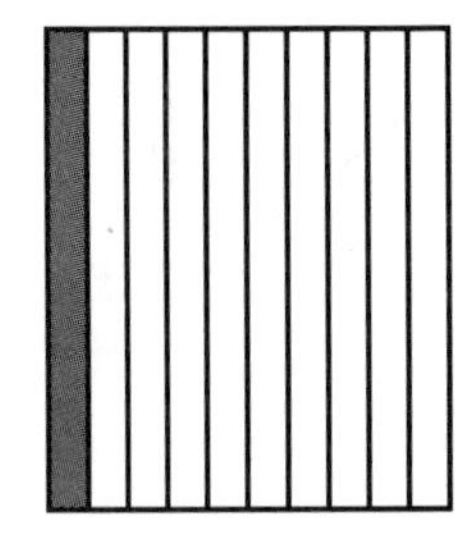
____

Write each fraction as a decimal.

7. $\frac{4}{10}$ ______ 8. $\frac{2}{10}$ ______ 9. $\frac{1}{10}$ ______ 10. $\frac{9}{10}$ ______ 11. $\frac{7}{10}$ ______

Write each decimal as a fraction.

12. 0.5 ____ 13. 0.3 ____ 14. 0.8 ____ 15. 0.6 ____ 16. 0.9 ____

Mixed Applications

17. A farmer has 2 sheep, 3 pigs, and 5 cows. Write a decimal and a fraction to show what part of the total number of animals is cows.

____

18. Tanya has 6 kittens. $\frac{1}{3}$ of the kittens are black. How many kittens are black?

____

Name ____________________

LESSON 23.2

Hundredths

Vocabulary

Fill in the blank.

1. There are 10 tenths in one whole. There are

   100 ____________ in one whole.

Shade the decimal square to show each amount. Write the decimal number that names the shaded part.

2. 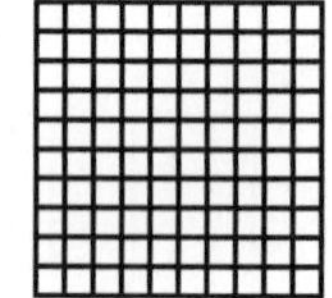

   seven hundredths

   ____

3. 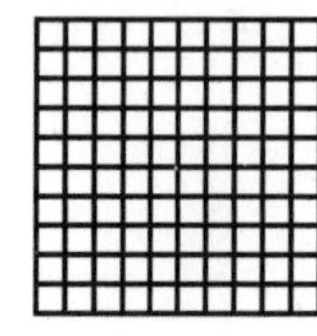

   nine hundredths

   ____

4. 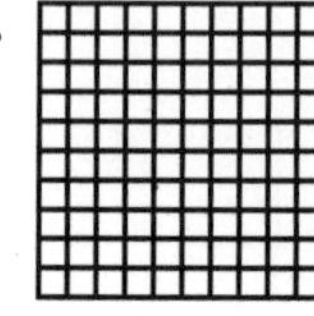

   twenty hundredths

   ____

5. 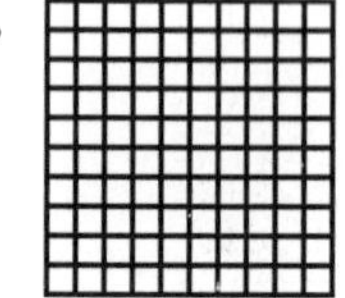

   twenty-five hundredths

   ____

6. 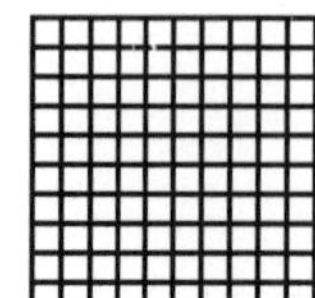

   forty-nine hundredths

   ____

7. 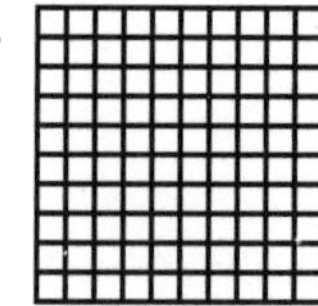

   seventy-two hundredths

   ____

For Exercises 8–9, use the tile design.

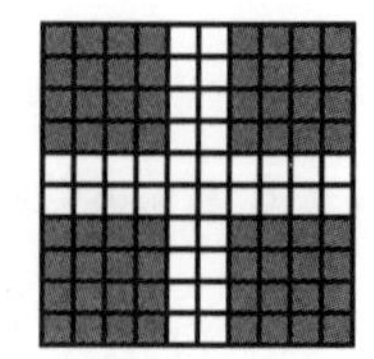

8. What decimal describes the gray tiles? ____________

9. What decimal describes the white tiles? ____________

Mixed Applications

10. Jack has 46 pennies. Write a decimal and a fraction that name the part of a dollar that Jack has.

    ____

11. Tammy has finished $\frac{3}{4}$ of her math problems. What part of her math problems has Tammy not yet finished?

    ____

Name ____________________

LESSON 23.3

Reading and Writing Hundredths

Vocabulary

Fill in the blank to complete the sentence.

1. A ____________________ separates the whole number from the fractional part of a number.

Record how you read and write the decimal name for each shaded part. The first example is done for you.

2. 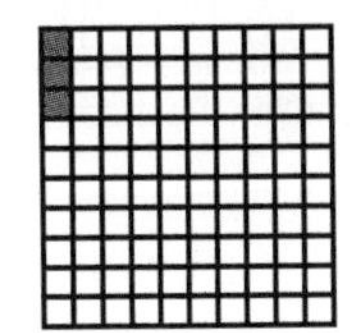

three hundredths;

0.03

3. 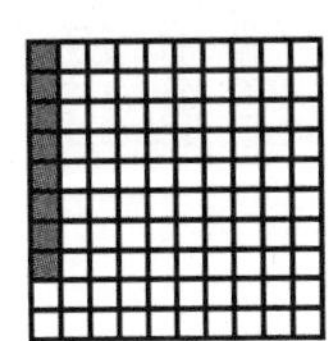

____

____

4. 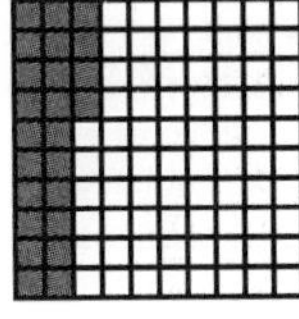

____

____

5. 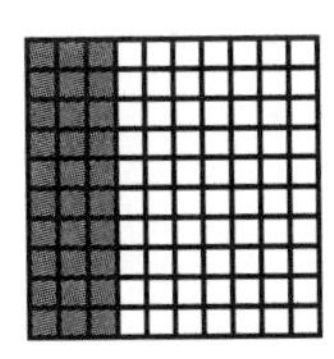

____

____

6. 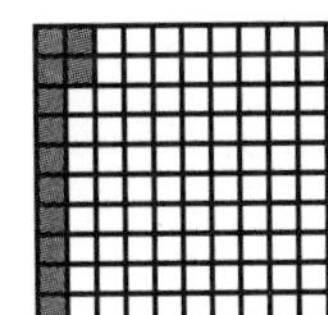

____

____

7. 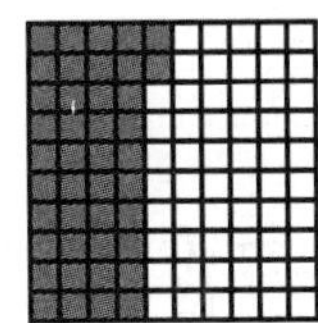

____

____

Mixed Applications

8. The Applegate Theater has 100 seats. On Saturday night, 78 seats were occupied. What decimal describes the seats that were not occupied? what fraction?

____

9. Tami rolled two number cubes that each have the numbers 1 to 6. The sum of the numbers she rolled was 8. List the possible pairs of numbers that Tami could have rolled.

____

Name ___________________________

LESSON 23.4

Decimals Greater Than 1

Vocabulary

Fill in the blank.

1. A ______________________ is a number that is made up of a whole number and a decimal.

Write the mixed decimal the model shows.

2. 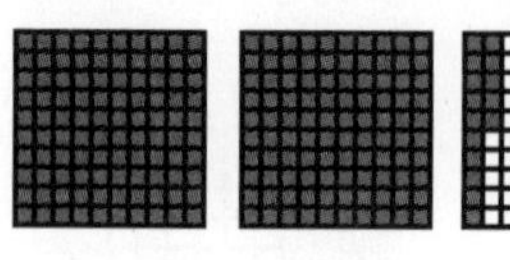
____

3. 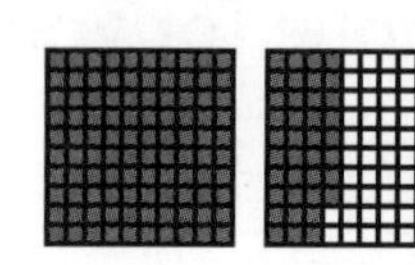
____

4. 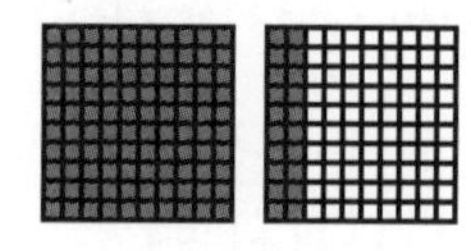
____

Write as a mixed decimal.

5. three and four tenths
____

6. five and thirty-four hundredths
____

Write each mixed decimal in words.

7. 4.3 ______________________

8. 6.25 ______________________

9. 24.1 ______________________

Mixed Applications

10. Each page of Bruce's album holds 10 stamps. Bruce has filled 6 pages. He has put 7 stamps on the next page. Write a mixed decimal to describe how many pages are full.
____

11. David buys 9 small marbles for 5¢ each, and he buys 5 large marbles for 9¢ each. What is his change from a $5 bill?
____

Name ______________________________

Comparing Decimal Numbers

Compare. Write < or > in each ○.

1. 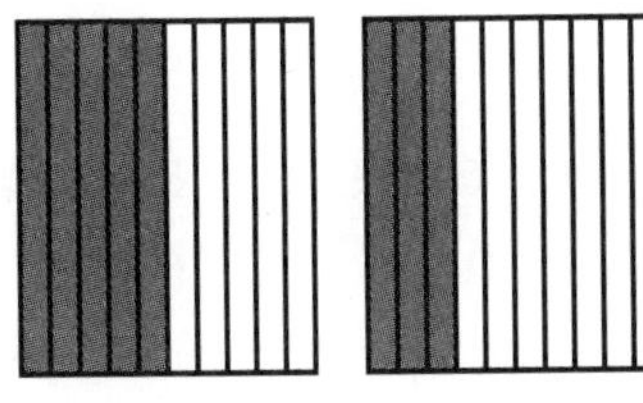 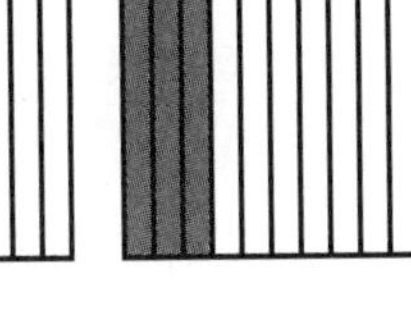

0.5 ○ 0.3

2. 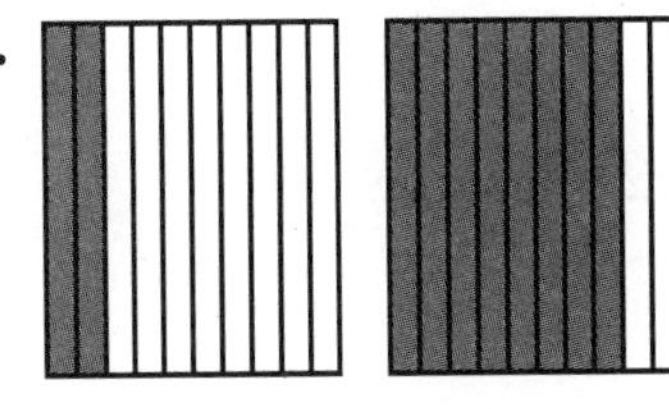

0.2 ○ 0.8

3. 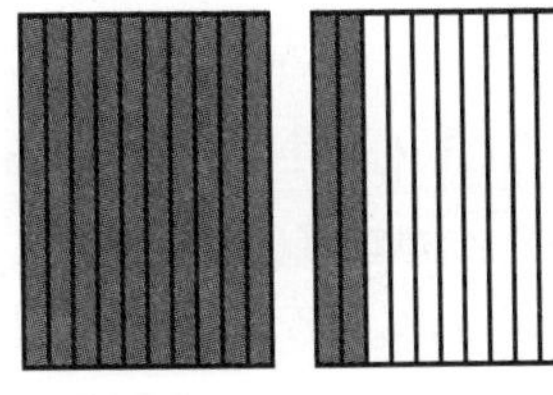 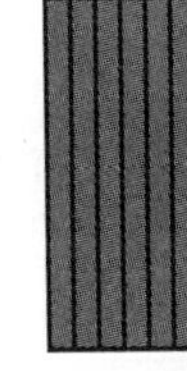 

1.2 ○ 1.4

4. 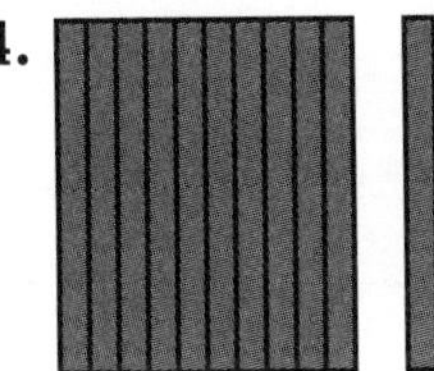 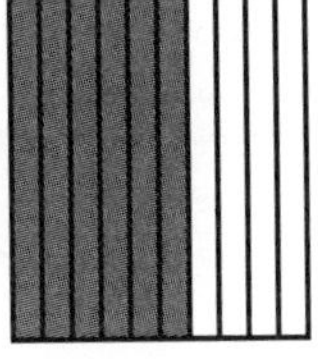 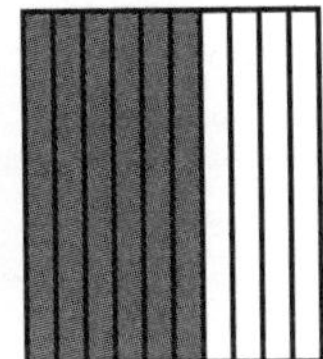 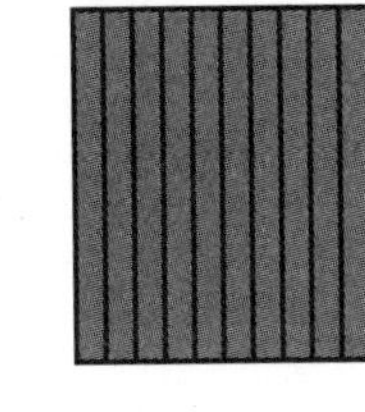

1.6 ○ 1.5

5.

Ones	Tenths
0.	4
0.	7

0.4 ○ 0.7

6.

Ones	Tenths
1.	4
1.	2

1.4 ○ 1.2

7.

Ones	Tenths
2.	8
2.	9

2.8 ○ 2.9

Mixed Applications

8. Sarah rode her bicycle 2.5 miles. Tim rode his bicycle 2.3 miles. Who rode farther?

____

9. Blair has three $1.00 bills and $0.56. How much more money does she need to buy a book that costs $5.00?

____

Name ______________________________

Problem-Solving Strategy

Draw a Picture

Draw a picture to solve.

1. Jessica is 1.2 meters tall. Dave is 1.4 meters tall. Who is shorter?

____

2. Jeff and Tom ordered a pizza that was cut into 8 slices. Jeff ate $\frac{3}{8}$ of the pizza, and Tom ate $\frac{5}{8}$. Who ate more pizza?

____

3. Mrs. Newman made 10 pancakes. Carla ate $\frac{4}{10}$ of the pancakes, and Pam ate $\frac{6}{10}$ of the pancakes. Who ate more pancakes?

____

4. Jo spelled 7 out of 10 words right on a spelling test. Tim spelled 0.8 of the words right on the same test. Who spelled more words right?

____

Mixed Applications

Solve.

CHOOSE A STRATEGY

- Draw a Picture
- Act It Out
- Make a Model
- Find a Pattern
- Write a Number Sentence

5. The sign in the bookstore says, *Sale! Buy 3 books and get 1 free.* Mrs. Jones wants to get 12 books. How many will she have to pay for?

____

6. There are 25 students standing in line in the cafeteria. There are 4 boys and 7 girls in front of Joe. How many students are behind Joe?

____

7. Tina is making a necklace of red and blue beads. She is repeating the pattern red, red, blue. What color will her fourteenth bead be?

____

8. The date is April 7. Jill's birthday will be in 2 weeks and 2 days. On what date is Jill's birthday?

____

Name ______________________

Inch, Foot, Yard, and Mile

Vocabulary

Fill in the blanks to complete the sentence.

1. The **inch (in.)**, **foot (ft)**, **yard (yd)**, and **mile (mi)** are customary units used to measure ______________ or ______________.

Choose the unit that you would use to measure each. Write *inch, foot, yard,* or *mile.*

2. the length of a table ______________

3. the length of a pinecone ______________

4. the length of a driveway ______________

5. the distance to a neighboring town ______________

Choose the better unit of measure. Write *inches, feet, yards,* or *miles.*

6. Your pencil is about 5 _____ long.

7. The distance from your home to the library is about 2 _____.

8. Your sled is about 4 _____ long.

9. The football player kicked the ball 45 _____.

10. Peter grew almost 2 _____ in one year.

11. Mr. Carlson is about 6 _____ tall.

Mixed Applications

12. You are making a shirt. You want to know exactly how long your arm is. What unit of measure would you use to measure the length of your arm? ______________

13. Bill is connecting 2-inch paper clips to make a chain. How many paper clips does Bill need to make a chain that is 14 inches long? ______________

Name ______________________________

LESSON 24.2

Estimating and Comparing Length

Estimate the length of each item. Then use a ruler to measure to the nearest inch.

1. 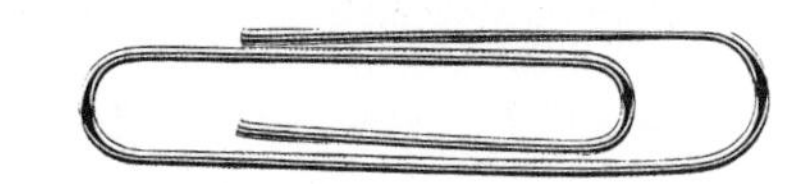

Estimate: ______________

2. 

Estimate: ______________

3. 

Estimate: ______________

4. 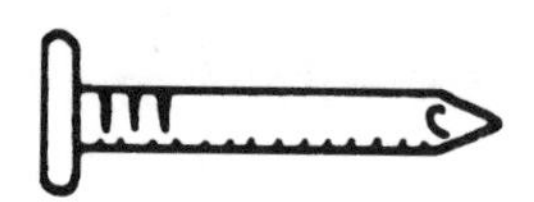

Estimate: ______________

5. 

Estimate: ______________

List three objects that measure about each given length.

6. 2 inches	7. 4 inches	8. 12 inches
______________	______________	______________
______________	______________	______________
______________	______________	______________

Mixed Applications

9. David estimated that his classroom is about 30 feet long. The actual measurement is 28 feet. What is the difference between David's estimate and the actual measurement?

____

10. Tanya buys pepper plants for her garden. She spaces them 18 inches apart in a straight row. How many inches from the first plant is the third plant?

____

Name ____________________

LESSON 24.3

Measuring to the Nearest Half Inch

Measure the length to the nearest half inch.

1. 

____

2. 

____

3. 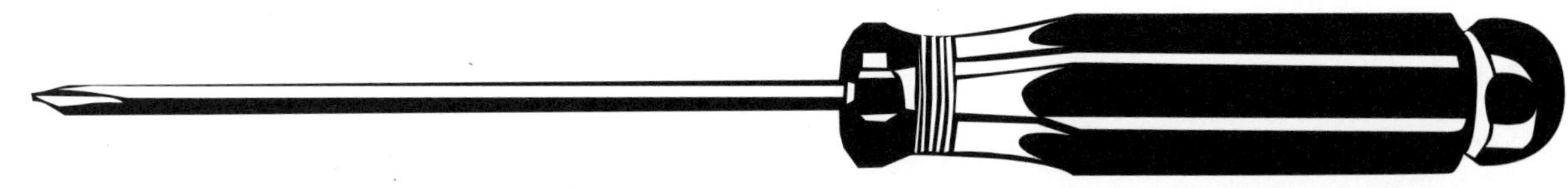

____

4. 

____

5. 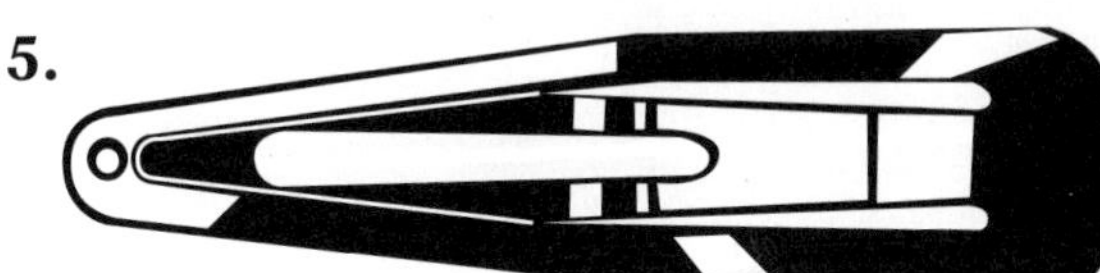

____

6. 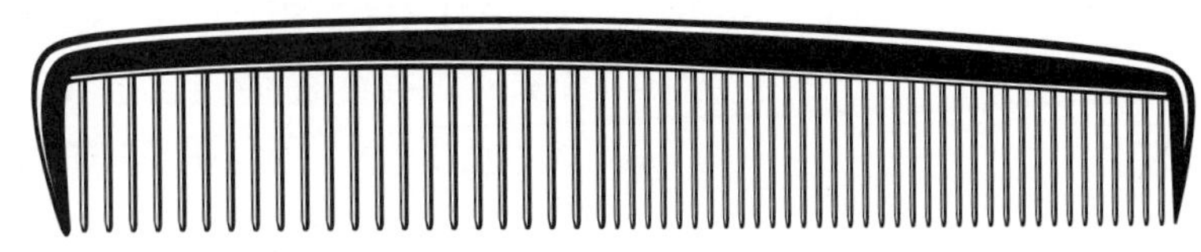

____

Mixed Applications

7. A plant measures $18\frac{1}{2}$ inches tall to the nearest half inch. Between which two inch marks does the top of the plant lie?

____

8. Alice has a 24-foot piece of rope. She cuts off 3 lengths of rope that are each 5 feet long. How long is the piece of rope that is left?

____

9. Tony glues a drawing to a piece of blue paper. The blue paper is 9 inches wide and 12 inches long. The blue paper makes a $\frac{1}{2}$ inch border around the drawing paper. How wide is the drawing paper?

____

10. Anita wants to hang a picture on the wall. The ceiling is 84 in. high. If she hangs the picture 16 in. below the ceiling, what is the height of the picture?

____

Name ______________________________

LESSON 24.3

Problem-Solving Strategy

Make a Model

Make a model to solve.

1. Suppose you want to make a bracelet that is 1 inch longer than the distance around your wrist. Measure your wrist to the nearest half inch. Tell how long your bracelet will be.

____

2. Rebecca has a scarf with a repeating pattern of red and white stripes. Each red stripe is 2 inches long, and each white stripe is 3 inches long. The total length of the scarf is 45 inches. How many red stripes does the scarf have?

____

3. Mr. Davis wants to cut a 12-foot board into 6 pieces that are each 2 feet long. How many cuts will he need to make?

____

4. Mary has 40-inch shoelaces. After lacing her shoes, she finds that there are 12 inches on each end for tying. How many inches are used to lace the shoes?

____

Mixed Applications

Solve.

CHOOSE A STRATEGY

- **Draw a Picture**
- **Act It Out**
- **Make a Model**
- **Find a Pattern**
- **Write a Number Sentence**

5. It is 4:00 when Sue and Jim begin their homework. Sue finishes in 45 minutes. Jim finishes 15 minutes before Sue. At what time does Jim finish his homework?

____

6. Barb practiced playing the flute 15 minutes on Monday, 18 minutes on Tuesday, and 21 minutes on Wednesday. If she continues this pattern, how many minutes will she practice on Friday?

____

Name ____________________

LESSON 24.4

Estimating and Comparing Capacity

Vocabulary

Complete.

1. ______________ is the amount of liquid a container can hold when filled.

2. Circle the words that are customary units for measuring capacity.

foot yard cup quart mile gallon inch pint

Circle the better estimate.

3. 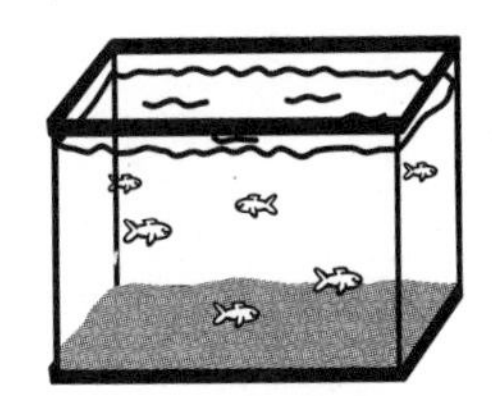

10 quarts or 10 gallons

4. 

2 cups or 2 quarts

Choose the unit that you would use to measure each. Write *cup, pint, quart,* or *gallon.*

5. milk in a glass ______________

6. punch for a party for 10 people ______________

7. water in a small pool ______________

Circle the greater amount.

8. 3 cups or 1 pint

9. 2 gallons or 9 quarts

10. 5 pints or 2 quarts

11. 1 gallon or 3 quarts

Mixed Applications

12. Rob pours 3 cups of milk from a full 1-quart container of milk. How much milk is left in the container?

____

13. Sam removes 9 links from a chain of 36 half-inch links. How long is the shortened chain? (You may *make a model* to solve.)

____

Name ____________________

Estimating and Comparing Weight

Vocabulary

Complete.

1. The customary units for measuring weight are ____________ and ____________. One pound equals 16 ounces.

Choose the unit that you would use to weigh each. Write *ounce* or *pound.*

2. 

____

3. 

____

4. 

____

5. 

____

6. 

____

7. 

____

Circle the better estimate.

8. 

4 ounces or
4 pounds

9. 

10 ounces or
10 pounds

10. 

10 ounces or
10 pounds

Mixed Applications

11. Leo is carrying 5 paperback books. Each book weighs about 4 ounces. Is Leo carrying more or less than a pound?

____

12. The grocery store is having a sale on bananas: 3 pounds for $1.00. Mrs. Lopez buys 6 pounds of bananas. What is her change from a $5 bill?

____

Name ______________________

Centimeter, Decimeter, Meter

Vocabulary

Circle the word that best completes each sentence.

1. The length of your arm is about 1 (centimeter/meter).
2. The width of your index finger is about 1 (decimeter/centimeter).
3. The width of an adult's hand is about 1 (decimeter/meter).

Choose the unit that you would use to measure each. Write *cm*, *dm*, or *m*.

4. a paper clip ______
5. the length of a classroom ______
6. a soccer ball ______
7. a flag ______
8. a ring ______
9. a car ______

Choose the unit that was used to measure each. Write *cm*, *dm*, or *m*.

10. An envelope is about 1 ______ wide.
11. A car is about 3 ______ long.
12. An adult's shoe is about 3 ______ long.
13. A tack is about 1 ______ long.
14. A window is about 2 ______ long.
15. A quarter is about 2 ______ wide.

Mixed Applications

16. You ran laps around the soccer field. Which unit of measure would you use to describe the distance you ran?

    ____

17. Todd has a music recital on June 30. As of today, he has four weeks to practice for the recital. What is today's date?

    ____

Name ______________________________

Estimating and Comparing Length

Estimate and measure each object. Use the unit of measure given.

1. the length of a sheet of construction paper (dm)

____

2. the length of a crayon (dm)

____

3. the length of a backpack (dm)

____

4. the height of 3 books (cm)

____

5. the length of a calculator (cm)

____

6. the length of a pencil's eraser (cm)

____

List two things that measure about as long as these measurements.

7. 2 centimeters

____

____

8. 2 decimeters

____

____

9. 2 meters

____

____

Mixed Applications

10. Beth cut an 80-centimeter tube into 2 pieces. One piece is 29 cm. How long is the other piece?

____

11. Mel has 7 dimes, 3 quarters, and 5 pennies. Does he have enough to buy a book that costs $1.50? Explain.

____

12. Carol has a wire that is 10 m long. She also has a 78 dm length of rope. Which is longer, the wire or the rope? Explain.

____

____

13. Jed is 8 dm tall. Bob is 83 cm tall. Who is taller? Explain.

____

____

Name ______________________

Measuring and Drawing Length

Measure each object or line.

1. 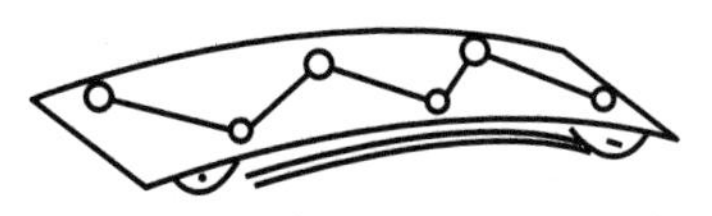

____

2. 

____

3. 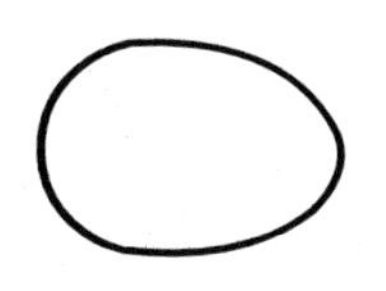

____

4. 

____

5.

____

6.

____

7. 

____

Draw a line of the given length.

8. 6 cm

9. 25 cm

10. 19 cm

Mixed Applications

11. One bug is about 4 cm long. How long is a chain of 5 bugs?

____

12. Carl's paper is 8 cm longer than Ann's paper. Ann's paper is 3 dm long. How long is Carl's paper?

____

13. Jill jogged 12 minutes longer than Ryan. They spent a total of 44 minutes jogging. How long did each person jog?

____

14. Eileen read 20 minutes longer than Joe. They read a total of 1 hour and 10 minutes. How long did each person read?

____

Name ______________________________

LESSON 25.3

Problem-Solving Strategy

Work Backward

Work backward to solve.

1. Mr. Ruiz sells mailboxes. He sold 5 mailboxes and then made 12 more. Now he has 15 mailboxes. How many did he begin with?

____

2. Paul has 23 outfielders and 19 pitchers in his baseball card collection. If he has a total of 95 cards, how many are not outfielders or pitchers?

____

3. Josh has 17 quarters and 28 dimes in his bank. There are 102 coins in the bank. How many are not quarters or dimes?

____

4. Tim sells picture frames. He sold 14 and then made 8 more. Now he has 23 frames. How many did he begin with?

____

Mixed Applications

Solve.

CHOOSE A STRATEGY

- **Draw a Picture**
- **Find a Pattern**
- **Make a Model**
- **Work Backward**

5. The dance recital is now over. It lasted for 2 hours. When did it start?

____

6. Which would cost less, 5 small erasers or 4 large erasers?

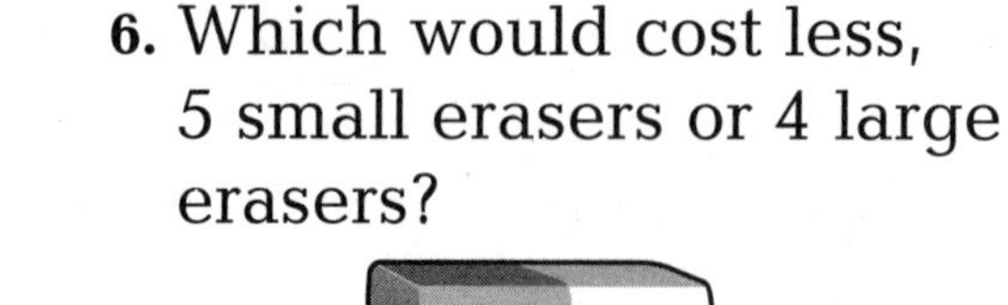

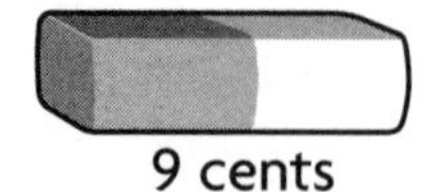

____

7. Rico jogged for 30 minutes the first week, 40 minutes the second week, and 50 minutes the third week. If this pattern continues, how long will he jog during the sixth week?

____

8. A store has 24 pairs of gloves for sale. There are 7 pairs of white gloves and 6 pairs of red gloves. How many pairs of gloves are not white or red?

____

Name ______________________________

LESSON 25.4

Estimating and Comparing Capacity

Vocabulary

Circle the word that best completes each sentence.

1. (Capacity/Meter) is the amount a container will hold when it is filled.
2. A water bottle has a capacity of 1 (milliliter/liter).
3. A medicine dropper has a capacity of 1 (milliliter/liter).

Circle the better estimate.

4. 

1 mL or 1 L

5. 

4 mL or 4 L

6. 

15 mL or 15 L

7. 

250 mL or 250 L

8. 

10 mL or 10 L

9. 

3,000 mL or 3,000 L

Choose the unit you would use to measure each. Write *mL* or *L*.

10. a mug of hot chocolate ________

11. water in a swimming pool ________

12. a glass of juice ________

Mixed Applications

13. A 12-L pot of sauce is divided equally into 3 smaller containers. How much sauce does each container hold?

____

14. Rod filled his dog's 400-mL water dish. Now the dish has 230 mL of water. How much did the dog drink?

____

15. Bud cut a 90-cm piece of rope into 3 equal pieces. How long is each piece?

____

16. Jon is 9 dm tall. Terri is 86 cm tall. Who is taller? Explain.

____

____

Name ____________________

LESSON 25.5

Estimating and Comparing Mass

Vocabulary

Circle the word that best completes each statement.

1. A book has a mass of about 1 (gram/ kilogram).
2. A dollar has a mass of about 1(gram/ kilogram).

Circle the better estimate.

3. 
6 g or 6 kg

4. 
25 g or 25 kg

5. 
22 g or 22 kg

6. 
4 g or 4 kg

7. 
6 g or 6 kg

8. 
2 g or 2 kg

Choose the unit you would use to measure each. Write *g* or *kg*.

9. a computer disk ________
10. a desk ________
11. a calculator ________
12. a pair of skates ________
13. a hair clip ________
14. a horse ________

Mixed Applications

15. The doctor is finding the mass of a young child. Will she use g or kg?

____

16. Box A has a mass of 2 kg. Box B has a mass of 450 g. Which box has a greater mass?

____

17. A 20-L container of juice is divided equally into 5 smaller jugs. How much does each jug hold?

____

18. Barry put 600 mL of water into a bird feeder. There is 225 mL remaining. How much water did the birds drink?

____

Name ______________________

LESSON 26.1

Finding Perimeter

Vocabulary

Fill in the blank to complete the sentence.

1. The distance around a plane figure is its ________________.

Find the perimeter of each figure.

2. 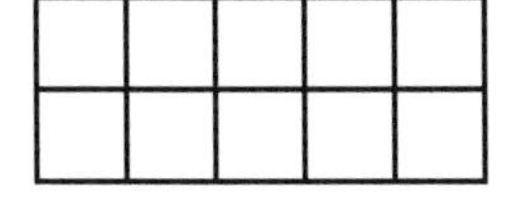

____

3. 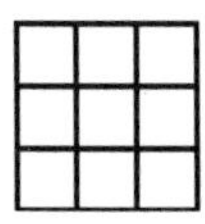

____

Use unit cubes. Find the perimeter of each figure.

4. 4 cm, 2 cm

____

5. 5 cm, 1 cm

____

6. 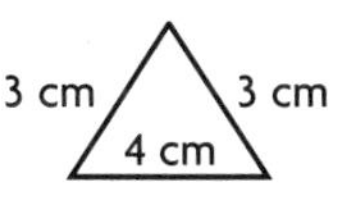

____

7. 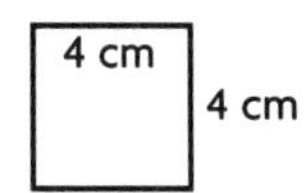

____

Mixed Applications

8. Patrick uses paper clips to measure his reading book. The book is 8 paper clips long and 5 paper clips wide. What is the perimeter of the book?

____

9. Lou left flyers on the door-steps of apartments 1, 4, 7, and 10. If he continues this pattern, which apartment is next?

____

10. In a group of 12 students, 3 students are wearing glasses. What fraction of the students are wearing glasses?

____

11. Julia is reading a book that has 218 pages. She has 25 pages left to read. How many pages has Julia read?

____

Name ______________________________

LESSON 26.2

More About Perimeter

Use your centimeter ruler to measure. Find the perimeter of each figure.

1. 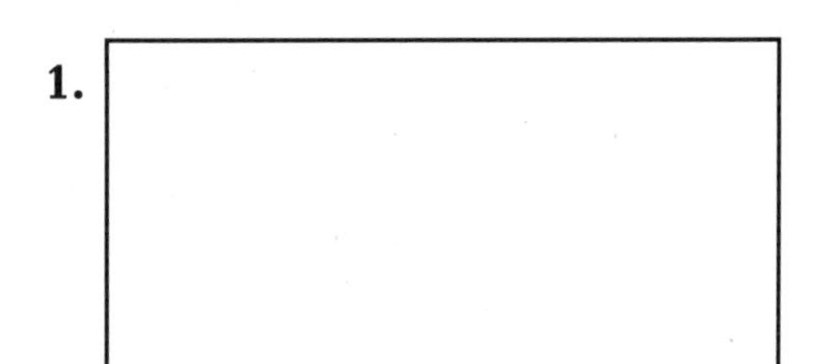

____

2. 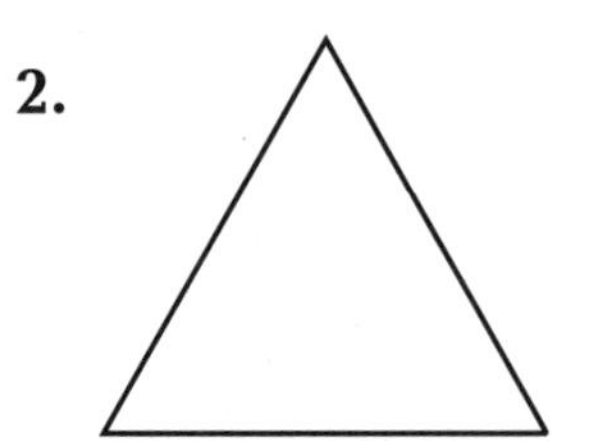

____

3. ______________

4. 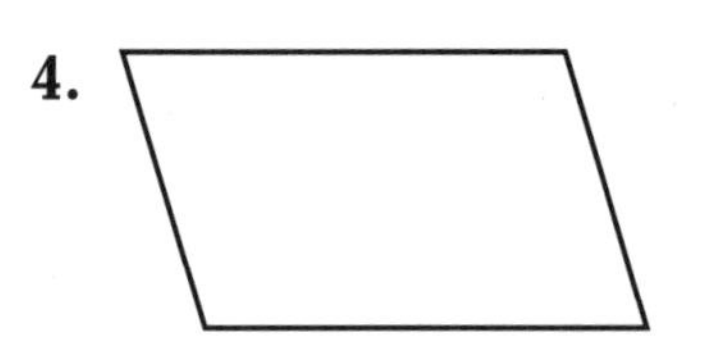

____

5. 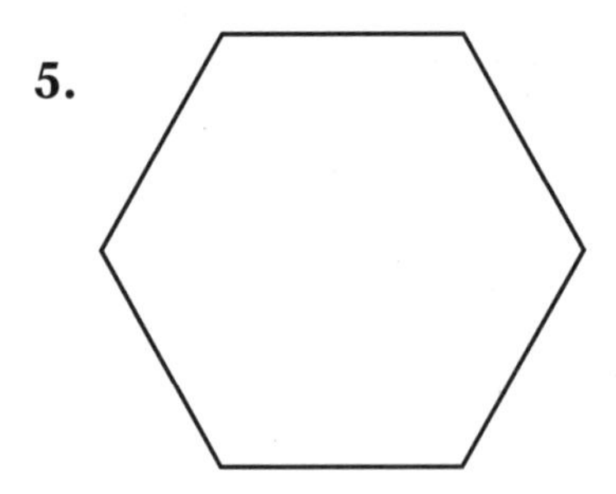

____

6. 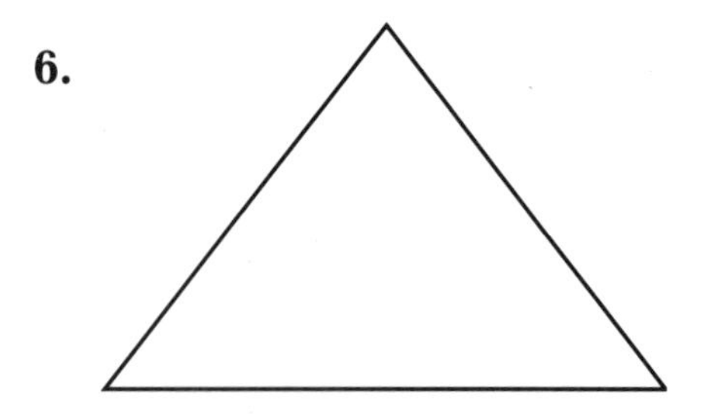

____

Mixed Applications

7. Mr. Jenkins is building a fence around a garden that is 3 yards wide and 9 yards long. How many yards of fence does Mr. Jenkins need?

____

8. Lilly left home with $6.00. Before she got back, she spent $2.95 and earned $4.50. How much did she have when she got back?

____

Name ______________________________

Finding Area

Vocabulary

Fill in the blanks to complete the sentence.

1. ____________ is the number of ____________ needed to cover a flat surface.

Use square tiles to make each figure. Draw the figures. Write the area in square units.

2. 2 rows of tiles, 4 tiles in each row

____

3. 2 rows of tiles, 3 tiles in each row

____

4. 4 rows of tiles, 1 tile in each row

____

5. 2 rows of tiles, 6 tiles in each row

____

6. 4 rows of tiles, 4 tiles in each row

____

7. 3 rows of tiles, 5 tiles in each row

____

Find the area of each figure. Label the answer in square units.

8. 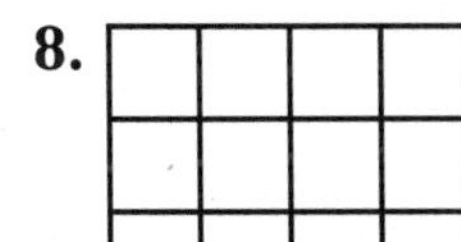

____

9. 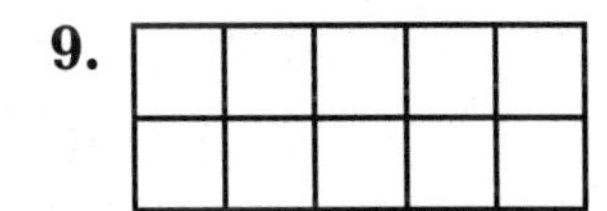

____

10. 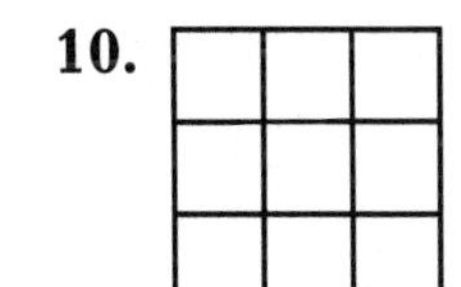

____

Mixed Applications

11. A tile floor has 7 rows with 5 tiles in each row. How many tiles are used to make the floor?

____

12. The perimeter of a square table is 12 feet. What is the length of each side of the table?

____

Name ______________________

LESSON 26.4

Perimeter and Area

Find the area and perimeter of the figure.

1. 4 ft, 2 ft

area: __________

perimeter: __________

2. 5 ft, 3 ft

area: __________

perimeter: __________

3. 4 ft, 1 ft

area: __________

perimeter: __________

4. 7 ft, 3 ft

area: __________

perimeter: __________

5. 9 ft, 2 ft

area: __________

perimeter: __________

6.

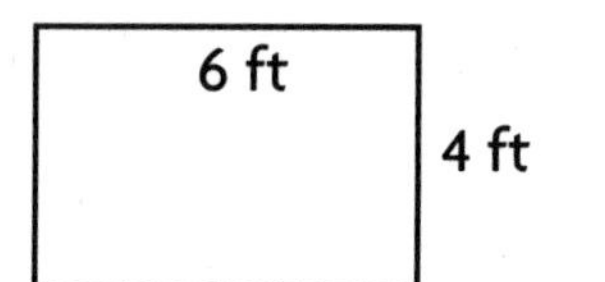

area: __________

perimeter: __________

Mixed Applications

7. Joe and Liz each have a garden plot with the same area. Joe's garden plot is 2 feet wide and 12 feet long. Liz's garden plot is 4 feet wide. What is the length of Liz's garden plot?

____

8. Tammy has a piece of ribbon that is 48 inches long. She uses the ribbon to make a border around a picture that is 12 inches long and 9 inches wide. How much ribbon is left over?

____

9. One strip of paper is 7 centimeters long. Another strip is 5 centimeters long. The two strips are taped together to make a strip that is 9 centimeters long. How long is the overlap?

____

10. A dog and a cat together weigh 45 pounds. The dog weighs 4 times as much as the cat. What is the weight of the cat? of the dog?

____

Name ______________________________

Problem-Solving Strategy

Act It Out

Act it out to solve.

1. A laundry room is shaped like a rectangle. The area of the room is 6 square yards. The perimeter is 10 yards. The room is longer than it is wide. How wide is the room? How long is the room?

____

2. Mark has a piece of string that is 12 inches long. He shapes the string into a rectangle that encloses an area of 5 square inches. Can Mark enclose a greater area with the same string? If so, what is the area?

____

3. The perimeter of a table is 24 feet. The table is twice as long as it is wide. What is the table's width? length? area?

____

____

4. Mrs. Brown put a wallpaper border around a room that is 10 feet long and 9 feet wide. How long is the wallpaper border? What is the area of the room?

____

Mixed Applications

Solve.

CHOOSE A STRATEGY

- **Draw a Picture** • **Act It Out** • **Make a Model** • **Work Backward** • **Write a Number Sentence**

5. The time shown on Mario's watch is 10:45. He has just finished raking leaves for 30 minutes. Before that, he played basketball for 1 hour. At what time did he start playing basketball?

____

6. Carrie is swimming in the middle lane of the pool. She waves to her father, who is swimming 3 lanes away, in the end lane. How many lanes does the pool have?

____

Name ______________________________

LESSON 27.1

Arrays with Tens and Ones

Put the tens and ones together. Name the factors of the new rectangle.

1. ______________

2. ______________

3. ______________

Use the array. Add the two products to find the answer. Complete the multiplication sentence.

4. 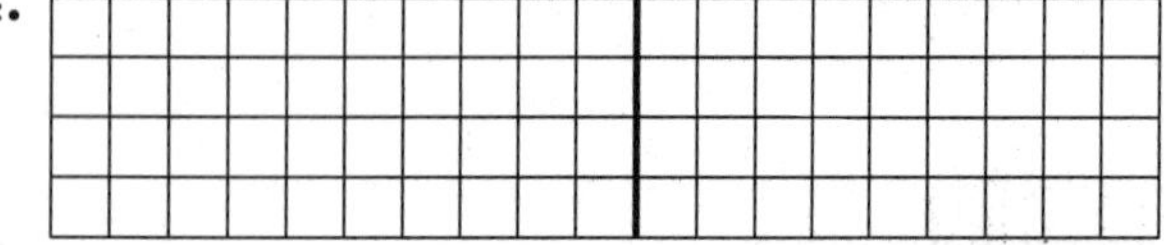

$4 \times 19 =$ ______________

____

5. 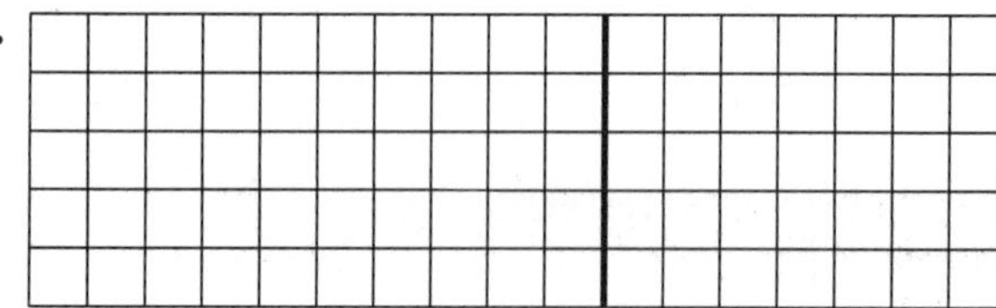

$5 \times 17 =$ ______________

____

Draw each array on grid paper. Show how you found the product.

6. $4 \times 12 = \underline{?}$

____

____

7. $3 \times 13 = \underline{?}$

____

____

8. $6 \times 14 = \underline{?}$

____

____

9. $5 \times 16 = \underline{?}$

____

____

Mixed Applications

10. Millie gave the cashier a $20.00 bill. She spent only $14.26. What was her change?

____

11. Henry made 14 cookies. Each cookie had 5 nuts. How many nuts in all did he use?

____

Name ______________________

Problem-Solving Strategy

Make a Model

Make a model to solve.

1. Sheila put all her stamps in a book. She has 6 rows of stamps, with 18 stamps in each row. How many stamps in all does Sheila have?

____

2. Annalee is making a quilt. The quilt will have 9 rows, with 12 squares in each row. How many squares will there be on Annalee's quilt?

____

3. A place mat has 8 rows of squares, with 15 squares in each row. How many squares are on the place mat?

____

4. Sal packs eggs into crates. He packed 6 rows, with 14 eggs in each row. How many eggs in all did Sal pack?

____

Mixed Applications

Solve.

CHOOSE A STRATEGY

- Find a Pattern
- Guess and Check
- Work Backward
- Write a Number Sentence
- Make a Model

5. Joyce bought passes to play miniature golf. The passes cost $3 for children and $5 for adults. Joyce spent $27. For how many adults and children did Joyce pay?

____

6. Steven is in a bike race. The race is 123 miles long. On Day 1 he rode 50 miles, and on Day 2 he rode 27 miles. How many miles does Steven have left to ride?

____

7. Brian sold 5 granola bars each day for 12 days. The granola bars cost $1.00 each. How many granola bars did he sell? How much money did he make?

____

8. Walter, Jeff, and Melanie are in line at the roller coaster. Jeff is behind Melanie, and Walter is in front of Jeff. Who is first, second, and third in line?

____

Name ___________________________

LESSON 27.2

Modeling Multiplication

Use base-ten blocks to find each product.

1. Each student in Kendra's class has 3 pencils. There are 23 students in her class. How many pencils in all do students in Kendra's class have?

   ____

2. There are 6 soccer teams in the spring tournament. Each team has 14 players. How many players are in the tournament?

   ____

3. $3 \times 15 = \underline{?}$ ___________

4. $4 \times 52 = \underline{?}$ ___________

5. $6 \times 32 = \underline{?}$ ___________

6. $5 \times 25 = \underline{?}$ ___________

7. $2 \times 46 = \underline{?}$ ___________

8. $5 \times 63 = \underline{?}$ ___________

9. $4 \times 29 = \underline{?}$ ___________

10. $3 \times 64 = \underline{?}$ ___________

11. $4 \times 67 = \underline{?}$ ___________

12. $3 \times 18 = \underline{?}$ ___________

13. $2 \times 51 = \underline{?}$ ___________

14. $6 \times 72 = \underline{?}$ ___________

Mixed Applications

15. On Sunday the temperature was 75°F. On Monday the temperature was 12° higher. What was the temperature on Monday?

    ____

16. Hugh sold 5 boxes of cookies. Each box contained 12 cookies. How many cookies in all did Hugh sell?

    ____

17. At the supermarket there were 4 rows of juice boxes. Each row had 25 boxes of juice. How many boxes of juice were there in all?

    ____

18. Stephanie watched television for 2 hours. She started watching television at 5:25. At what time did she stop watching television?

    ____

Name ____________________

LESSON 27.3

Recording Multiplication

Find each product. Use base-ten blocks to help you.

1. 56×4
2. 29×2
3. 64×3
4. 24×5
5. 13×4
6. 84×5
7. 45×7
8. 36×8
9. 24×2
10. 32×6
11. 47×7
12. 29×4
13. 18×3
14. 51×2
15. 27×4
16. 33×6

Mixed Applications

17. Christine and Julie went shopping. They spent $20 on CDs, $30 on jeans, and $15 on food. They started with $77. How much money do they have left?

____

18. Carlos returned to the doctor's office on May 17. This was exactly 2 weeks after he sprained his wrist. What was the date Carlos sprained his wrist?

____

19. Together, Celia and Kyle washed 36 dishes. Kyle washed 6 more dishes than Celia. How many dishes did each of them wash?

____

____

20. Mrs. Kay had her students sit in rows of 5. There are 6 students in each row. How many students are there in Mrs. Kay's class?

____

Name ______________________________

LESSON 27.4

Practicing Multiplication

Find the product. Use base-ten blocks to help you.

1. 96×3 = ____
2. 21×2 = ____
3. 83×5 = ____
4. 56×6 = ____
5. 71×3 = ____
6. 45×2 = ____
7. 69×5 = ____
8. 83×3 = ____

9. In which of Exercises 1–8 did you need to regroup?

____

10. 75×3 = ____
11. 28×7 = ____
12. 16×4 = ____
13. 33×2 = ____

14. $84 \times 2 =$ ____
15. $64 \times 3 =$ ____
16. $32 \times 5 =$ ____

Mixed Applications

17. Sarah bought 5 tulip plants. Each plant cost $4. How much in all did the plants cost?

____

18. José walked 3 miles every day for 14 days. How many miles in all did José walk?

____

19. Eileen and Charlie had a party. They invited 12 guests. Each guest received 3 party favors. How many party favors did Eileen and Charlie buy?

____

20. Mark earned $36 mowing lawns. He mowed 6 lawns in all. How much was he paid to mow each lawn?

____

Name ______________________________

LESSON 28.1

Dividing with Remainders

Vocabulary

Fill in the blank to complete the sentence.

1. In division, the ______________ is the amount left over.

Use counters to find the quotient and remainder.

2. $13 \div 3 =$ _____ 3. $15 \div 2 =$ _____ 4. $11 \div 4 =$ _____

5. $12 \div 5 =$ _____ 6. $10 \div 4 =$ _____ 7. $9 \div 5 =$ _____

Use the model to find the quotient and remainder.

8. $17 \div 3 =$ _____

Step 1 | Step 2

_____ counters

Quotient is _____.

Remainder is _____.

9. $13 \div 4 =$ _____

Step 1 | Step 2

_____ counters

Quotient is _____.

Remainder is _____.

Find the quotient and remainder. You may use counters to help you.

10. $23 \div 4 =$ _____ 11. $30 \div 4 =$ _____ 12. $25 \div 3 =$ _____

Mixed Applications

13. Marissa baked 18 cookies. She put the same number of cookies into 4 different bags and ate the cookies that were left over. How many cookies did she put in each bag?

____

14. All of the students in Mrs. Lee's class are working in small groups on art projects. There are 8 groups of 3 students and one group of 2 students. How many students are there in the class?

____

Name ______________________________

LESSON 28.2

Modeling Division

Use the model to find the quotient and remainder.

1. $51 \div 2 =$?

Step 1

_____ tens _____ ones

Step 2

Quotient is _____.

2. $38 \div 3 =$?

Step 1

_____ tens _____ ones

Step 2

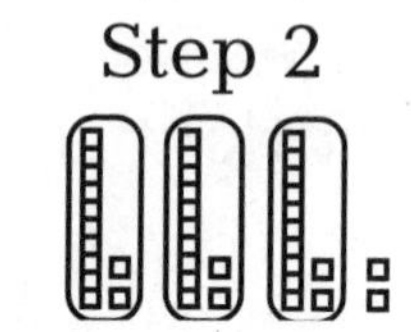

Quotient is _____.

Find the quotient. Use base-ten blocks to help you.

3. $53 \div 2 =$ ________
4. $61 \div 4 =$ ________
5. $17 \div 2 =$ ________
6. $63 \div 5 =$ ________
7. $48 \div 5 =$ ________
8. $48 \div 3 =$ ________

Mixed Applications

9. There are 56 campers traveling in 4 vans. The same number of campers are riding in each van. How many campers are there in each van?

____

10. Carlos has finished solving 8 out of 10 math problems. What fraction of the problems has Carlos not solved yet?

____

11. One carton holds 24 cans of soup. How many cans of soup are there in 4 cartons?

____

12. Sarah cut a rope 5 times to divide it into 6-inch long pieces. How long was the rope before she cut it?

____

Name ______________________________

Recording Division

Find the quotient. You may wish to use base-ten blocks. Then check each answer.

1. $35 \div 3 = ?$ Check: $3\overline{)35}$ So, $35 \div 3 =$ ____	2. $31 \div 2 = ?$ Check: $2\overline{)31}$ So, $31 \div 2 =$ ____	3. $49 \div 4 = ?$ Check: $4\overline{)49}$ So, $49 \div 4 =$ ____
4. $27 \div 5 = ?$ Check: $5\overline{)27}$ So, $27 \div 5 =$ ____	5. $48 \div 3 = ?$ Check: $3\overline{)48}$ So, $48 \div 3 =$ ____	6. $65 \div 4 = ?$ Check: $4\overline{)65}$ So, $65 \div 4 =$ ____

Mixed Applications

7. Mary picked 64 flowers. She divided them evenly in 4 vases. How many flowers did she put in each vase?

____

8. Brian works at the recycling center. He earns $8 an hour. How much money does he earn in a 35-hour workweek?

____

Name ______________________________

LESSON 28.4

Practicing Division

Find the quotient, using only paper and pencil. Check each answer by using multiplication.

1. $29 \div 4 =$ ______ Check: $4\overline{)29}$	2. $67 \div 5 =$ ______ Check: $5\overline{)67}$	3. $63 \div 4 =$ ______ Check: $4\overline{)63}$
4. $56 \div 3 =$ ______ Check: $3\overline{)56}$	5. $39 \div 2 =$ ______ Check: $2\overline{)39}$	6. $51 \div 3 =$ ______ Check: $3\overline{)51}$

Mixed Applications

7. Jeff, Chris, and Carrie are sharing a bag of 80 marbles. How many marbles does each person get? How many marbles are left over?

____

____

8. Troy paid for a sandwich with a $5 bill. The clerk gave him two $1 bills,1 dime, and 1 nickel in change. How much did the sandwich cost?

____

Name ____________________

Choosing Multiplication or Division

Write whether you should multiply or divide. Solve each problem.

1. Susan's family paid $96 for 4 used bicycles. Each bicycle cost the same amount. How much did each bicycle cost?

   ____

2. A third-grade class learns 18 new spelling words each week. How many words does the class learn in 3 weeks?

   ____

3. A lunch room seats 84 students. Each table can seat 6 students. How many tables are in the lunch room?

   ____

4. Maria has written 24 pages in her diary. She puts 3 daily entries on each page. How many daily entries has she written?

   ____

Mixed Applications

For Problems 5–7, use the table.

WASHINGTON SCHOOL ENROLLMENT	
Grade	**Number of Students**
Third	56
Fourth	55
Fifth	52

5. The third-grade classes are going to a museum. They divide into 4 equal groups to look at the exhibits. How many students are in each group?

   ____

6. The third- and fourth-grade classes both have recess at 10:30. How many students in all have recess at 10:30?

   ____

7. There are 28 boys in the fifth grade. How many of the fifth graders are girls?

   ____

8. Jeremy uses 60 feet of fencing to go all the way around a square garden plot. What is the length of each side of the garden plot?

   ____

9. Today is March 15. If Anne's kittens are born today, on what date will they be 2 weeks old?

   ____

Name ______________________________

LESSON 28.5

Problem-Solving Strategy

Write a Number Sentence

Write a number sentence to solve.

1. Each table in the cafeteria can seat 6 students. How many tables are needed to seat 96 students?

____

____

2. A parking lot has 8 rows of spaces, with 16 spaces in each row. How many cars can the lot hold?

____

____

3. Last year, Chelsea spent $46 on camera film. This year she spent 3 times that amount. How much did she spend on film this year?

____

____

4. Rob had 26 watermelon seeds. He planted 3 seeds in each cup. How many cups did he use? How many seeds were left over?

____

____

Mixed Applications

Solve.

CHOOSE A STRATEGY

- **Act It Out**
- **Draw a Picture**
- **Use a Table**
- **Guess and Check**
- **Write a Number Sentence**

5. Sue spends 50 minutes reading and writing. She spends 20 more minutes reading than writing. How many minutes does she spend reading?

____

6. Jed needs 32 inches of ribbon to make a frame around his picture. The picture is 9 inches long. How wide is the picture?

____

7. Mr. Frank cuts a large pizza into 8 slices. He sells 3 of the slices. What fraction of the pizza is left?

____

8. Tammy wants to buy skates that cost $29.00. She has saved $15.65. How much more money does she need?

____